Introduction to Emerging Fields in Materials Sustainability

Other World Scientific Titles by Seeram Ramakrishna

An Introduction to Biocomposites
ISBN: 978-1-86094-425-3
ISBN: 978-1-86094-426-0 (pbk)

An Introduction to Electrospinning and Nanofibers
ISBN: 978-981-256-415-3
ISBN: 978-981-256-454-2 (pbk)

Polymer Membranes in Biotechnology: Preparation, Functionalization and Application
ISBN: 978-1-84816-379-9
ISBN: 978-1-84816-380-5 (pbk)

The Changing Face of Innovation: Is it Shifting to Asia?
ISBN: 978-981-4291-58-3 (pbk)

Additive Manufacturing: Foundation Knowledge for the Beginners
ISBN: 978-981-12-2481-2
ISBN: 978-981-12-2624-3 (pbk)

An Introduction to Biomaterials Science and Engineering
ISBN: 978-981-12-2817-9

Sustainability for Beginners: Introduction and Business Prospects
ISBN: 978-981-12-4193-2
ISBN: 978-981-12-4316-5 (pbk)

Introduction to Emerging Fields in Materials Sustainability

Pankaj Pathak

SRM University Andhra Pradesh, India

Susmita Sharma

National Institute of Technology Meghalaya, India

Ramadoss Tamil Selvan

National University of Singapore, Singapore

Seeram Ramakrishna

National University of Singapore, Singapore

World Scientific

NEW JERSEY · LONDON · SINGAPORE · BEIJING · SHANGHAI · HONG KONG · TAIPEI · CHENNAI · TOKYO

Published by

World Scientific Publishing Co. Pte. Ltd.

5 Toh Tuck Link, Singapore 596224

USA office: 27 Warren Street, Suite 401-402, Hackensack, NJ 07601

UK office: 57 Shelton Street, Covent Garden, London WC2H 9HE

Library of Congress Cataloging-in-Publication Data
Names: Pathak, Pankaj, author. | Sharma, Susmita, author. |
    Tamil Selvan, Ramadoss author. | Ramakrishna, Seeram, author.
Title: Introduction to emerging fields in materials sustainability / Pankaj Pathak,
    Susmita Sharma, Ramadoss Tamil Selvan, Seeram Ramakrishna.
Description: New Jersey : World Scientific, 2024. | Includes bibliographical references and index.
Identifiers: LCCN 2023042120 | ISBN 9789811247644 (hardcover) |
    ISBN 9789811247651 (ebook) | ISBN 9789811247668 (ebook other)
Subjects: LCSH: Refuse and refuse disposal. | Circular economy. | Sustainable development.
Classification: LCC TD793 .P38 2024 | DDC 628--dc23/eng/20240304
LC record available at https://lccn.loc.gov/2023042120

British Library Cataloguing-in-Publication Data
A catalogue record for this book is available from the British Library.

For any available supplementary material, please visit
https://www.worldscientific.com/worldscibooks/10.1142/12568#t=suppl

Desk Editor: Shaun Tan Yi Jie

Typeset by Stallion Press
Email: enquiries@stallionpress.com

PREFACE

This book highlights the need for sustainability in a modern-day society where concepts like circular economy and zero waste are gaining traction. A comprehensive overview of various aspects of waste generated from different origins and causes is described here. Cases/best practices from countries which have migrated from linear economy and are implementing the concept of circular economy in different types of waste, *viz.* e-waste, food waste, plastics, C&D waste, munipical solid waste, etc., are discussed. The book presents the latest research perspectives, technology advances, and critical thinking developments that will be an essential guide to understanding the global waste crisis and help provide pragmatic viewpoints that inspire students, researchers, policymakers, and organizations to facilitate the move toward a zero waste economy.

Chapter 1 explains the fundamentals and context of sustainability in general. It presents the global policies and schemes for environmental sustainability and the action plan needed for carbon neutrality. The perspective of a modern sustainable society is projected.

Chapter 2 reviews current practices for Sustainable Development Goals and sustainability. Emerging fields focusing on sustainability are introduced.

Chapter 3 focuses on the global initiative for e-waste management. It also highlights the various barriers that have been identified as slowing the implementation of material recovery through e-waste recycling.

Chapter 4 can be used to provoke more sustainable thinking and integrate sustainability principles in our entire food supply chain process to reduce and prevent food waste, via the life cycle assessment (LCA) method and circular economy principle.

Chapter 5 addresses the challenges associated with the mismanagement of plastic waste, a potential threat to the terrestrial and marine environments. It discusses how biobased plastics represent an alternative to petro-based plastics.

Chapter 6 considers methods for selecting the best construction and demolition (C&D) waste management. Alternative legislation requiring waste generators or owners to classify, recycle and reuse C&D are examined.

Chapter 7 showcases the landmark policies and initiatives on municipal solid waste management and global trends of solid waste generation. It provides an overview of integrated solid waste management based on the waste management hierarchy to reduce waste disposed of while maximizing resource recovery and efficiency.

Chapter 8 describes the characteristics and properties of natural waters and the need for water sustainability. It also focuses on the advances in treatment technologies for wastewater recycling.

Chapter 9 depicts the application of nanomaterials for fostering sustainability in different fields such as biomedicine, wastewater treatment, energy storage devices, etc., underscoring the multidisciplinary nature of this topic.

Chapter 10 summarizes the life cycle thinking of any product or service and implementing a circular economy. It explains how LCA can assess the process and required resources, i.e., input water, energy, and materials.

Through its comprehensive exploration of sustainability practices, policy frameworks, innovative technologies, and global best practices, this book serves as a critical resource for understanding the challenges and opportunities in modern waste management. We hope to inspire a paradigm shift — from reactive waste disposal to proactive resource recovery — and ultimately guide readers toward building a more circular, resilient, and zero-waste future.

Pankaj Pathak
Susmita Sharma
Tamil Selvan
Seeram Ramakrishna

ABOUT THE AUTHORS

Dr. Pankaj Pathak is an Associate Professor in the Department of Environmental Science and Engineering at SRM University Andhra Pradesh, India. She holds a PhD from IIT Bombay, India. Her research area includes sustainable waste management, green energy resources, techno-economic analysis, and life cycle assessment. She has acquired research experience in dealing with hazardous solid and liquid waste and its treatment technologies, along with the recovery of critical and rare earth metals from waste streams, their safe disposal, and remediation techniques.

Dr. Susmita Sharma is an Assistant Professor in Department of Civil Engineering at National Institute of Technology Meghalaya, India. She obtained her PhD in Geotechnical Engineering from IIT Bombay, India. She received an opportunity to work as a Visiting Researcher at the University of Strathclyde, Scotland, by the support of European Commission via the Marie Curie

IRSES project GREAT. She specializes in Environmental Geotechnology, with particular emphasis on utilization of anthropogenic sediments of wastewater treatment plants. Currently, her primary areas of research include waste material characterization for contamination evaluation and possible methods of management.

Mr. Ramadoss Tamil Selvan is a Research Engineer at the National University of Singapore, Singapore. Before this appointment, he was working as a Chief Technical Officer in startup companies. He has been actively involved in sustainability business development and prospects by working with startup companies. He is a keen entrepreneur, innovator, speaker, and developed sustainability/circularity software tool for industries and academics. He received an MSc from the University of Nottingham, UK, and a BEng from the Anna University, India. He also received a Nottingham Business School scholarship to pursue an MBA.

Professor Seeram Ramakrishna, *FREng* is the Chair of Circular Economy Taskforce at the National University of Singapore (NUS). Microsoft Academic ranked him among the top 36 salient authors out of three million materials researchers worldwide (https://academic.microsoft.com/authors/192562407). He is the Editor-in-Chief of the Springer NATURE Journal Materials Circular Economy — Sustainability (https://www.springer.com/journal/42824). He is a member of Enterprise Singapore's and International Standards Organization's Committees on ISO/TC323 Circular Economy and Circularity. He teaches ME6501 Materials and Sustainability module (https://www.europeanbusinessreview.com/

circular-economysustainability-and-business-opportunities/), and also mentors Integrated Sustainable Design ISD5102 course projects on reimagining and enhancing circularity of Industrial Estates. He is an opinion contributor to the Springer Nature Sustainability Community (https://sustainabilitycommunity.springernature.com/users/98825-see ramramakrishna/posts/looking-through-covid-19-lens-for-a-sustainable-new-modern-society). He is an advisor to the National Environmental Agency's CESS events (https://www.cleanenvirosummit.sg/programme/ speakers/professor-seeramramakrishna; https://bit.ly/catalyst2019video; https://youtube.com/watch?v=ptSh_1Bgl1g). He is an Impact Speaker at University of Toronto, Canada's Low Carbon Renewable Materials Center (http://www.lcrmc.com/#). He is a judge for the Mohammed Bin Rashid Initiative for the Global Prosperity (https://www.facebook. com/Make4Prosperity/videos/innovation-inclusivetrade/47950353933 9143/). He is an advisor to the Singapore company TRIA (www.triabio 24.com) which specializes in zero waste packaging.

Professor Seeram, with a h-index of 161 as of May 2021, is named among the World's Most Influential Minds (Thomson Reuters), and the Top 1% Highly Cited Researchers (Clarivate Analytics) (https://scholar. google.com.sg/citations?hl=en&user=a49NVmkAAAAJ&view_op=list_ wor ks&sortby=pubdate). He is an elected Fellow of UK Royal Academy of Engineering (FREng); Singapore Academy of Engineering; Indian National Academy of Engineering; and ASEAN Academy of Engineering & Technology. He is appointed as the Honorary Everest Chair of MBUST, Nepal. His senior academic leadership roles include NUS University Vice-President (Research Strategy); Dean of NUS Faculty of Engineering; Director of NUS Enterprise; and Founding Chairman of Solar Energy Institute of Singapore (http://www.seris.nus.edu.sg/). He received PhD from the University of Cambridge, UK; and TGMP from the Harvard University, USA.

CONTENTS

CHAPTER ONE
BASICS OF SUSTAINABILITY AND ITS DIMENSIONS

1.1 What is Sustainability?

Sustainability is a broad term which has been defined in many ways (see right panel). The concept is not new and has been a social practice in many parts of the globe since early civilization. Nature conservation, exploitation of natural reserves in a limited manner, and even worshipping natural sources (e.g., water bodies, forests, suns, sea, etc.) were an integral part of ancient civilizations worldwide. With time, to achieve rapid growth and fulfil demands to make our lives more comfortable, we started to

SUSTAINABILITY

- "Sustainability is to meet the needs of the present without compromising future generation's ability to meet their own needs." — *World Commission on Environment and Development*
- "Sustainability is a lifestyle designed for performance." — *Chris Turner*
- "We must aim for a continuous, resilient, and sustainable use (of forests)." — *Hans Carl von Carlowitz*
- "A desire to create a society that is safe, stable, prosperous, and ecologically minded." — *Jeremy L. Caradonna*
- "The ability of a system to endure and be maintained at a certain rate over time." — *The authors of this book.*

exploit our resources in an unlimited manner despite knowing that they are finite.

The term "sustainability" in its modern context was introduced by the World Commission on Environment and Development in 1987. It was then that Norwegian Prime Minister Gro Harlem Brundtland defined sustainability in the report *Our Common Future* as: *"meeting the needs of the present without compromising the ability of future generations to meet their own needs."*[1] It is important to clarify that although "sustainability" is often associated primarily with climate change, it is actually a broader and more complex concept. The term encompasses not only environmental concerns but also social well-being, human health, industrial development, and economic stability.

Therefore, sustainability should not be considered solely a synonym for environmental protection. Its meaning depends on the objectives and actions of individuals, groups, and communities working collaboratively toward common goals. Over time, the definition of sustainability has evolved, and several key events have marked milestones in its development, as shown in Table 1.1. Previously, the United Nations identified four domains of sustainability: economic, environmental, political, and cultural. However, in the post-COVID-19 era, this framework has been updated to include resource management, environmental protection, social well-being, knowledge integration, and the circular economy.[2]

Despite these developments, the environment, society, and economy remain the core elements — commonly referred to as the three pillars of sustainability (Figure 1.1a). This concept is influenced by the field of ecology, which emphasizes the dynamic interactions among human communities, resource flows, and the natural environment. On the other hand, there is another view that these three components do not hold equal weight. From this perspective, the environment forms the foundational layer, as society cannot exist — nor can it sustain economic

Table 1.1   The milestones of a sustainable world.

Year	Event
1979	First World Climate Conference opens up the science of climate change
1987	Brundtland Report consolidates decades of work on sustainable development and defines sustainability
1992	Rio Earth Summit rallies the world to take action and adopt Agenda 21
1993	Convention on Biological Diversity puts the precautionary principle to work
1997	Kyoto Protocol takes the first step toward stopping dangerous climate change
2000	With Millennium Development Goals, social justice meets public health and environmentalism
2002	IPCC introduces carbon neutrality for achieving sustainability
2007	IPCC recommends minimizing global temperature rise to 2°C
2012	Rio+20 takes stock of two decades of efforts at sustainable development
2015	United Nations summit on sustainable development proposes 17 goals for sustainable development
2020	United Nations sets the objectives for sustainable development "For People For Planet"

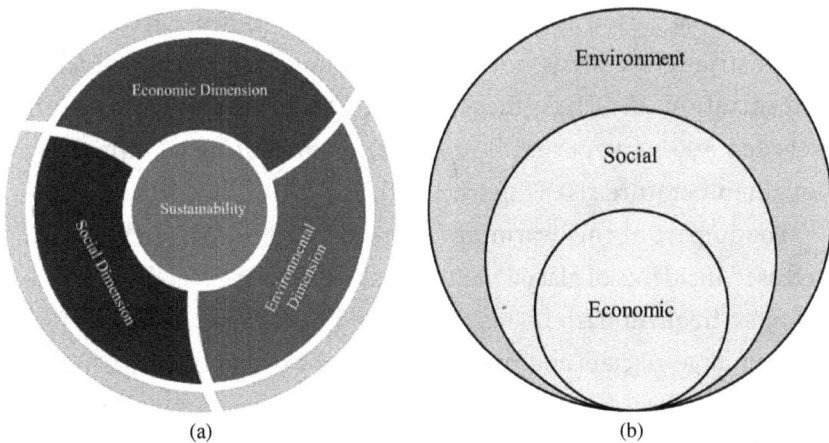

(a)   (b)

Figure 1.1   (a) Model illustrating equal weightage among the three core components of sustainability — environment, society, and economy. (b) Environmental foundation model, emphasizing the environment as the fundamental base supporting social and economic systems within the sustainability framework.

activity — without a healthy environment. This model of sustainability is illustrated by placing the environment as the base (Figure 1.1b).

1.2 Pillars of Sustainability

1.2.1 Environmental Sustainability

The environment is one of the most crucial pillars of sustainability and is often associated with critical global issues such as climate change, global warming, and the emission of greenhouse gases (GHGs). These environmental challenges are increasingly evident today, as we experience hotter temperatures on a daily basis.

According to recent data from the National Oceanic and Atmospheric Administration,[3] the year 2024 was the tenth-warmest year on record, with global surface temperatures 2.32°F (1.29°C) above the 20th-century average of 57.0°F (13.9°C) and 2.63°F (1.46°C) warmer than the pre-industrial era (1850–1900). Using the pre-industrial period as a baseline, Figure 1.2a shows that our pattern of industrial growth has not been environmentally sustainable. The concentration of GHGs, measured in CO_2-equivalent (CO_2e), has exceeded 400 parts per million (ppm), contributing significantly to global temperature rise (Figure 1.2b).

The impact of this warming is already visible. We are witnessing increased incidents of glacial melt, global sea level rise, biodiversity loss, and more frequent flash floods across the globe. Current estimates suggest that large glaciers and polar ice caps are melting at rates 2–3 times faster than in the previous century. These developments indicate that our historical trajectory of growth has been largely unsustainable from an environmental perspective.

As the long-term consequences of exponential industrial growth and energy consumption become increasingly evident, it is imperative that we take action to reverse these effects and prevent further environmental

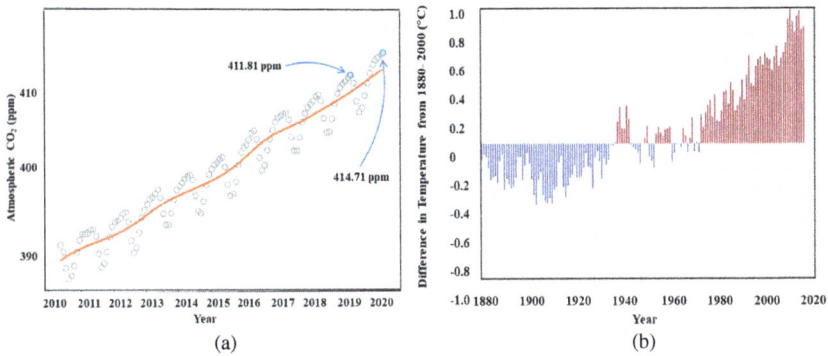

Figure 1.2 (a) Global average temperature represented by the Keeling curve. Reproduced with permission from https://grist.org/climate/flatten-the-curve-coronavirus-climate-emissions/. (b) Yearly average surface temperature from 1880–2022 — blue bars indicate cooler-than-average years; red bars show warmer-than-average years. Reproduced with permission from https://www.climate.gov/news-features/understanding-climate/climate-change-global-temperature.

damage, ensuring that future generations inherit a healthy and livable planet. Environmental sustainability focuses on protecting and conserving natural resources — such as water, air, glaciers, forests, mountains, minerals, and even overall cleanliness — for the benefit and use of future generations. To support this goal, several key aspects that contribute to building resilient communities and securing sustainable growth are illustrated in Figure 1.3.

It is important to recognize that the environmental impact of our actions or decisions is not always immediately visible. Therefore, a core principle of environmental sustainability is its forward-looking nature. In simple terms, it emphasizes the idea: "Act today to secure the future." This is why many environmental regulations and policies are already in place, and it is essential that we comply with them to protect the environment not just for today, but for the generations to come. These regulations include setting and enforcing standards for the quality of soil, water, air, wildlife habitats, and greenhouse gas emissions. Penalties and legal actions are also in practice to ensure adherence to sustainable and

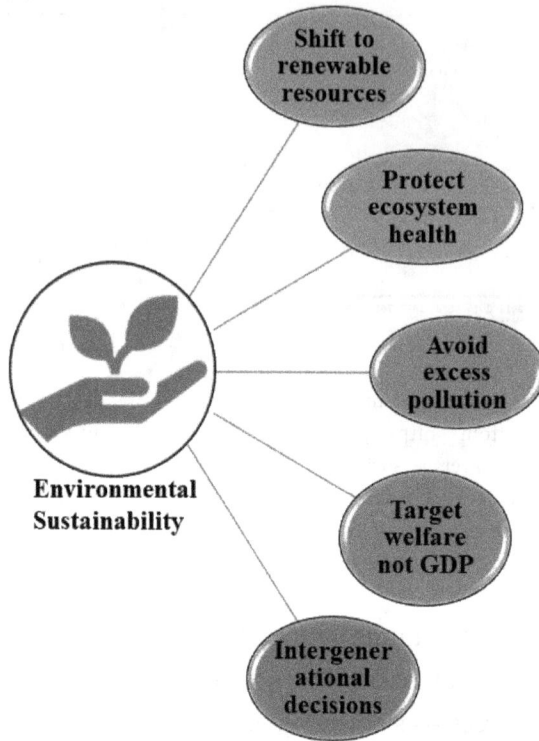

Figure 1.3 Different aspects of environmental sustainability.

responsible environmental behaviors. A summary of some globally promoted environmental policies and initiatives is presented in Table 1.2.

1.2.1.1 *Carbon Neutrality*

In the context of environmental sustainability, the terms "carbon neutrality" and "net-zero emissions" are frequently used and highly sought after. According to the United Nations, carbon neutrality refers to removing from the atmosphere an amount of CO_2 equivalent to what an organization has emitted. On the other hand, net zero represents a broader and more ambitious goal — it involves not only achieving

Table 1.2   Global policies and schemes for environmental sustainability.

Policies and Schemes	Overview
Carbon tax	A tax placed on the production/consumption of carbon, e.g., burning fossil fuels. The aim is to make users face the total social cost instead of just the private cost.
Carbon finance	Covers financial tools such as carbon emission trading to reduce the impact of GHGs on the environment by giving carbon emissions a price.
Carbon credit	A permit that allows the company that holds it to emit a certain amount of CO_2 or other GHGs. One credit permits emission equal to one ton of CO_2.
Carbon offset	Buying of credits to compensate for the GHG emissions released by industries or entities.
Carbon trading	The process of buying and selling permits and credits to emit CO_2.
Carbon footprint	Measures the amount of CO_2 produced (in tons) as a result of various anthropogenic activities.
Low carbon/ decarbonized economy	An economy that uses low-carbon power sources with minimal emission of GHGs into the atmosphere.
Clean development mechanism	In 1997, under the Kyoto Protocol (Annex B), a commitment was made to implement an emission reduction project in developing countries.
Principle of responsible investment	Introduced environmental, social, and corporate governance (ESG) issues that affect the performance of investment portfolios (to varying companies, sectors, regions, asset classes, and through time).

carbon neutrality but also eliminating or offsetting all types of GHG emissions produced by an organization (Table 1.3).[4]

In short, carbon neutrality implies attaining net zero GHG or carbon dioxide equivalent (CO_2e) emissions, and this can be accomplished by balancing or eliminating those emissions through the planet's natural absorption (wetland restoration, reforestation, etc.) or carbon offsetting/ mechanical (carbon capture technologies, green energy infrastructure, etc.) via climate action. Generally, carbon neutrality defines a means of

Table 1.3　Action plan for carbon neutrality and net zero.

Definition	Action Plan
Carbon Neutral: Removes an equal amount of CO_2 as emitted.	• Climate pledge made by the management team. • Driven by evolving legislation.
Net Zero: Removes an equal amount of all the GHG or CO_2e as emitted.	• Leverage carbon markets to offset emissions. • Strive for emissions footprint reduction.

large-scale production where the total output of CO_2e is neutral, i.e., equal to zero. This does not mean that companies will have zero emissions, but these emissions are offset, i.e., counterbalanced. In this context, computation of CO_2e necessitates an emission factor (EF), which is the amount of CO_2e emitted per unit use of the product:

$$CO_2e = EF \times \text{quantity used} \tag{1.1}$$

Carbon neutrality was entered into force by 195 countries at the 2015 United Nations Climate Change Conference (COP21) in Paris. The Paris Agreement set out the target to limit global warming to well below 2°C by the end of the century and pursue efforts to limit it further to 1.5°C and minimize carbon emissions. The emissions reduction target is part of the global path to becoming carbon neutral by 2050. According to ISO 14067 or ISO 14064, the Carbon Neutral Protocol outlines the steps for an organization to obtain carbon-neutral certification. Generally, two approaches are implemented to calculate carbon footprint: (1) the organization, and (2) the product. The Carbon Neutral Protocol helps to achieve the SDG 13 (climate action) and SDG 9 (industry, innovation, and infrastructure). The United Nations has recommended implementing ESG practices to achieve more sustainability and carbon neutrality by 2050.

The top seven emitters (China, the United States, India, the European Union, Indonesia, Russia, and Brazil) accounted for about half of global GHG emissions in 2020. The Group of 20 (Argentina, Australia, Brazil, Canada, China, France, Germany, India, Indonesia, Italy, Japan, Republic of Korea,

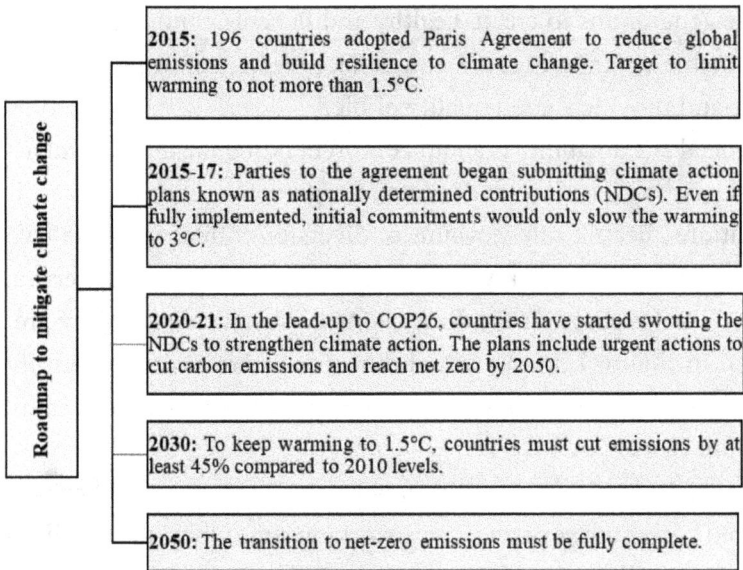

Figure 1.4 A roadmap to mitigate carbon emissions.

Mexico, Russia, Saudi Arabia, South Africa, Turkey, the United Kingdom, the United States, and the European Union) are responsible for ~75% of global GHG emissions. Therefore, it is of utmost importance for these nations to take immediate action to mitigate carbon emissions. Accordingly, around 400 organizations from sports to aviation have taken a pledge to become carbon neutral by 2050. The roadmap to net zero carbon emissions from 2015 to 2050 has been proposed and is shown in Figure 1.4.

1.2.2 Social Sustainability

Sustainable development discussions often spotlight the environmental or economic aspects of sustainability, but overlook social sustainability. As per the Western Australia Council of Social Services, "Social sustainability occurs when the formal and informal processes, systems, structures, and relationships actively support the capacity of current and future generations to create healthy and liveable communities. Socially sustainable communities are equitable, diverse, connected, and democratic and provide a good quality of life."

Social sustainability is about comprehending businesses' impacts on people and society. It includes human rights, fair labor practices, living conditions, health, safety, wellness, diversity, equity, work-life balance, empowerment, community engagement, philanthropy, volunteerism, and more. The four significant dimensions of social sustainability are presented in Figure 1.5a; the social dimensions are easier to identify but issues associated with these dimensions are not very easy to quantify or measure (Figure 1.5b).

Indeed, social sustainability is a business investment. When workers are paid fairly and work under safe working conditions, they are healthier and more productive. This gives more profits to companies. Therefore, to assess the contribution of social sustainability toward sustainable development, we need to know how to measure our progress over time using social indicators. As a result, a range of methodologies and tools have been developed for assessing social indicators for sustainability achievement across products and technologies. In this context, the United Nations Environment Programme presented the life cycle sustainability

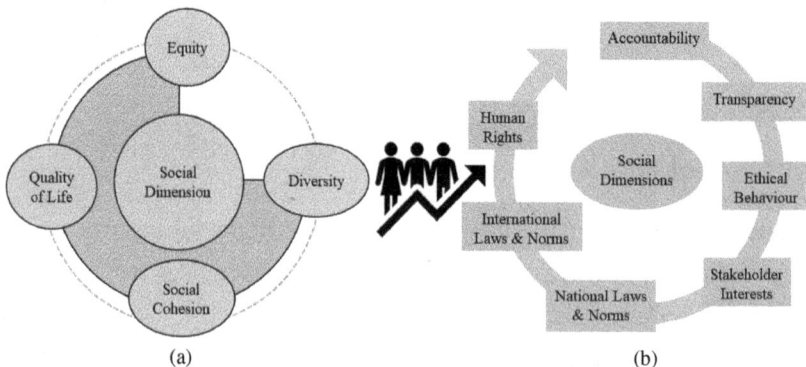

Figure 1.5 (a) Significant dimensions of social sustainability. (b) Issues associated with social sustainability.

assessment that accounts for social sustainability. Further, ISO 26000 introduced social responsibility in the life cycle assessment (LCA) model as the willingness of a corporation/organization to incorporate social and environmental considerations in its decision making. It is termed social life cycle assessment (S-LCA) and aims to assess products' socio-economic impacts along with their life cycle, from "cradle to grave". The work provides a roadmap to carry out the entire analysis in four main frameworks, as shown in Figure 1.6.

In the S-LCA, relevance analysis is conducted to assess the data collection efforts of the social aspects. Briefly, the social compatibility (SC) and unit process relevance (UPR) are multiplied to obtain a social relevance number (SRN). The SRN has a very similar principle to risk failure mode and effects analysis in risk management.

$$SRN = SC \times UPR \tag{1.2}$$

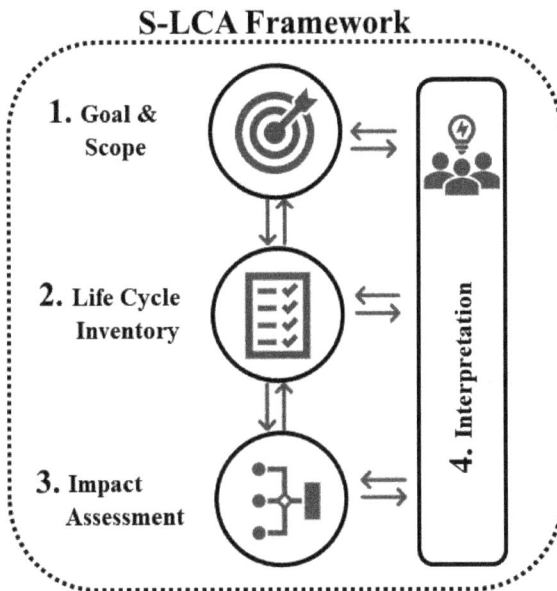

Figure 1.6   Framework for S-LCA.

Table 1.4 Description of SHDB ranking, social compatibility, unit process relevance and social relevance number.

SHDB Ranking	Social Compatibility	Social Compatibility Ranking	Relevance in Life Cycle	Unit Process Ranking	Social Relevance Number
No risk	Very high	0	—	—	0
Low	High	1	Low	1	2
Medium	Limited	2	Medium	2	6
High	Low	3	High	3	12
Very high	Very low	4	Very high	4	16
No data	Unknown	N/A	—	—	—

The first step for conducting the social relevance analysis is to evaluate the information obtained from the social hotspot database (SHDB). The data in the SHDB describes the risk assessments ranging from 0 to 4, where 0 is evidence of no risk, 1 is low, 2 is medium, 3 is high and 4 is very high. If there is no data available sector-specific or country-specific, it shows as not applicable or no data. A low SHDB rating describes high SC and a very high rating shows very low SC, as shown in Table 1.4. Further, life cycle relevance is quantified by UPR. The maximal values of SC and UPR are 4; hence when SRN is calculated the maximum SRN comes to 16 but, generally, threshold relevance for SRN lies between 4 and 6.

1.2.3 *Economic Sustainability*

Economic sustainability is an integrated part of sustainability and means that we must use, safeguard and sustain resources (human and material) to create long-term sustainable values by optimal use, recovery, and recycling.

However, there is a massive debate about to what extent economic growth conflicts with environmental sustainability. On one side, rising gross domestic product and output lead to a higher consumption, greater pollution, and greater demand for natural resources. Garret Hardin observed that in unregulated markets, overgrazing or overfishing occurred that lead to market failure and a decline in environmental sustainability. This is known as the "Tragedy of the Commons", where damage occurs due to innocent actions by individuals.

However, inevitable economic growth can still be compatible with environmental sustainability via technological development and utilizing renewable resources. Therefore, it is not mandatory to stop economic growth, but rather, adapt economic growth by focusing on environmental sustainability. In this context, the Kuznets curve is a controversial theory stating that the environment degrades at one stage of economic development and after a turning point, a post-industrial service sector leads to declining environmental degradation.

1.2.3.1 Circular Economy

Given the current environmental challenges, the traditional cradle-to-grave model — commonly described as Take → Make → Consume → Dispose — is increasingly recognized as unsustainable. This linear approach leads products to a fixed end-of-life stage, often resulting in large amounts of waste.

To address this, a cradle-to-cradle approach has been proposed as a sustainable

> **CIRCULAR ECONOMY**
>
> - "An economy which is restorative and regenerative by design, and aims to keep products, components, and materials as their highest utility and value at all times." — *Ellen MacArthur Foundation*
>
> - "An industrial system that replaces the end-of-life concepts with restoration, shifts towards the use of renewable energy, eliminates the use of toxic chemicals, which impair reuse and return to the biosphere, and aims for the elimination of waste through the superior design of materials, products, systems, and business models." — *The World Economic Forum*
>
> - "A regional economic model in the era of the 5Rs which replaces the input of imported energy and virgin materials with the skilled labor of local workers, optimizing the use of existing resources and stocks." — *Patil and Ramkrishna*

alternative. In this model, products are not discarded after use; instead, post-consumer materials are recovered through regeneration, rejuvenation, or resynthesis, allowing them to be reused in the manufacturing of new products. This minimizes the need for virgin raw materials and fosters a closed-loop system, as illustrated in Figure 1.7.

To support this shift, the Circular Economy has emerged as a key operational framework designed to bridge the gap between production systems and natural ecological cycles. It is built on three core principles:

1. Redesign systems to eliminate waste and pollution,
2. Reuse products and materials wherever possible,
3. Regenerate natural systems to maintain ecological balance.

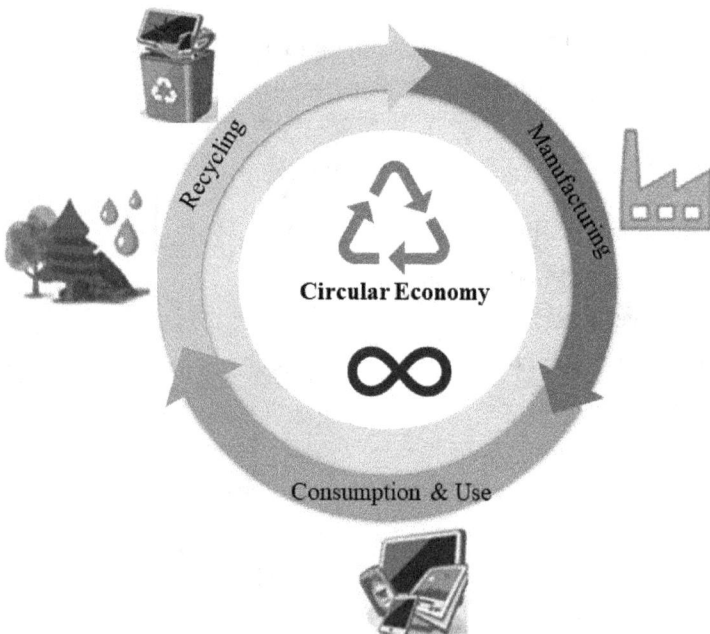

Figure 1.7 A circular economy model illustrating the cradle-to-cradle manufacturing approach, designed to mitigate the limitations of the traditional linear model.

In recent developments, the circular economy has expanded to incorporate additional multi-R strategies, including reduction (in consumption) and recovery (in terms of regeneration and recycling efficiency). Furthermore, the circular economy can be viewed through four strategic perspectives:

1. Green and sustainable product development,
2. Business model or supply chain integration,
3. Green marketing and sustainable consumption, and
4. Hybrid model construction and optimization.

Through these dimensions, the circular economy demonstrates its potential to enhance sustainability, with measurable impacts in economic and monetary terms, while contributing to the mitigation of environmental degradation and resource depletion.[5]

1.3 Sustainable Development Goals

In response to historically unsustainable growth patterns — particularly those driven by linear economic activities — the concept of sustainable development was introduced during the United Nations Conference on Environment and Development in 1992. This concept emerged as a strategy to address global sustainability challenges and promote development that does not rely on the continuous depletion of natural resources. A core principle of sustainable development is the decoupling of economic growth from the overuse of natural resources, and aiming to establish achievable and inclusive goals for all people, regardless of their local, national, regional, or international context. Within this framework, two distinct modes of goal-setting have been introduced.

1.3.1 Millennium Development Goals for Sustainability

At the start of the 21st century, the United Nations Millennium Declaration was announced, which commits to a global alliance to reduce extreme poverty and embark on a series of eight time-bound goals: (1) eradicate extreme poverty; (2) achieve universal primary education; (3) promote gender equality and empower women; (4) reduce child mortality; (5) improve maternal health; (6) combat HIV/AIDS, malaria and other diseases; (7) ensure environmental sustainability; and (8) create a global partnership for development. For achieving the goals in developing nations, the deadline was set 15 years away, in 2015, and it was known as the Millennium Development Goals (MDGs). The MDGs report optimistically stated that it was the most successful anti-poverty program in history. MDGs, in retrospect, supplemented the visions of the Asian Development Bank, that is, the region should be free of poverty and citizens' quality of life in developing and underdeveloped nations should be improved.

1.3.2 United Nations Goals for Sustainable Development

The efforts to achieve sustainable growth under the MDGs took pace, which consequently led the world leaders of 193 countries to to come together in 2015 to establish a new set of global guidelines. This collaboration resulted in the creation of the Sustainable Development Goals (SDGs) — a framework of 17 interlinked objectives designed to serve as a shared blueprint for peace, prosperity, and sustainability for the planet, the environment, and humanity, both now and in the future. They are often referred to as the 17 Goals of SDGs (see Figure 1.8), envisioning to achieve by 2030 No poverty — SDG 1, Zero hunger — SDG 2, Good health and well-being — SDG 3, Quality education — SDG 4, Gender equality — SDG 5, Clean water and sanitation — SDG 6, Affordable and clean energy — SDG 7, Decent work and economic growth — SDG 8, Industry, innovation, and infrastructure — SDG 9, Reduced inequalities — SDG 10, Sustainable cities and communities — SDG 11, Responsible consumption and production — SDG 12, Climate action — SDG 13, Life below water — SDG 14, Life on land — SDG 15,

Figure 1.8   United Nations goals for sustainable development.

Peace, justice, and strong institutions — SDG 16, Partnerships for the goals — SDG 17. These goals are often cross-cutting and synergistic with each other (e.g., SDG 13 on climate change is synergistic with SDG nos. 3, 7, 11, 12, and 14).

1.4 Perspective for a Modern Sustainable Society

A multi-perspective approach encourages awareness and change to build a sustainable path to modern society. The different expressions shown in Figure 1.9 exhibit the prime objectives of sustainable cities and communities. The different colors adopted for the expressions indicate their nature in different sectors.

SDG 11 is designated to sustainable cities and communities and underpins the role of the individual in sustainable use of natural resources and reducing environmental degradation. A sustainable modern society exercises the urban mining concept. Urban mining is the process where

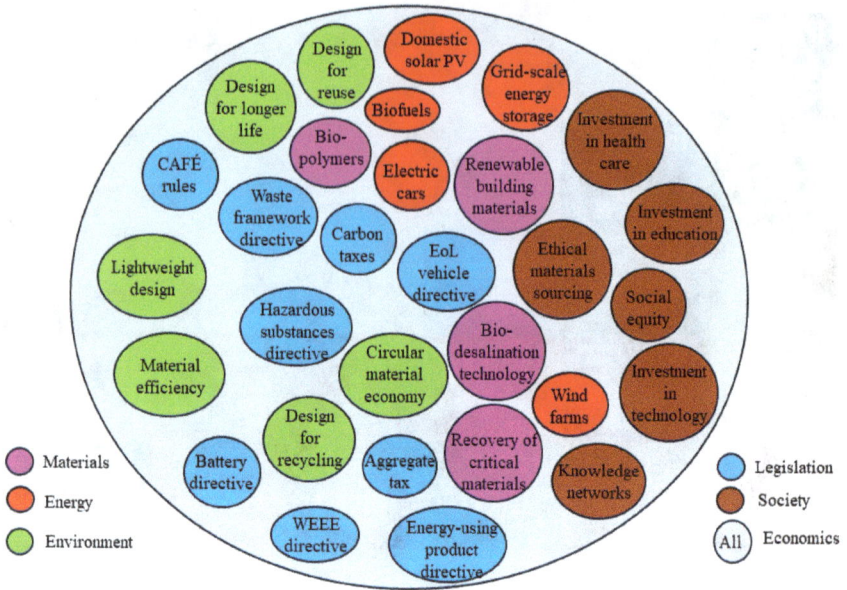

Figure 1.9   The prime objectives of the sustainable modern society and their impact on all six sectors.

waste generated from cities/communities such as municipal solid waste, e-waste, construction waste, etc., would be reclaimed as a secondary resource. Urban mining aims to monetize waste streams in such a way as to create revenue, business, and jobs. Here, it not only minimizes the environmental burden but also boosts the economy.

In this view, a global sustainable city index (SCI), as shown in Table 1.5, is developed by the United Nations Environment Programme. The 12 indicators of SCI show positive environmental externalities and fulfill the 17 SDGs by minimizing the ecological and carbon footprints of the city/community. The values for each of the 12 indicators are converted into a score out of 1.0.[6]

Furthermore, the Organization for Economic Co-operation and Development has proposed five evaluation indicators in building a smart and green city as shown in Figure 1.10.

Table 1.5 A benchmark for green and smart city index (Source: Sustainable City Index Report, 2023).[6]

	SCI Indicators	Target	Unit	Average for Cities in HIC	Average for Cities in LIC	Weightage (%)
1	GHG emissions	0	tonnes CO_2e/cap	4.5	2.3	10
2	Consumption-based emissions	0	tonnes CO_2e/cap	17.1	6.4	10
3	Particulate air pollution (PM2.5)	<5	$\mu g/m^3$	10.5	27.8	20
4	Open public space	45	%	12	14.5	5
5	Water access	100	%	100	95	5
6	Water consumption	100–150	L/cap/day	254	182	5
7	Road infrastructure efficiency	<1	km/km^2	3.9	2.2	5
8	Sustainable transport mode share	75	%	42	60	5
9	Automobile dependence	<1	vehicles/hh	1.0	0.7	5
10	Solid waste generated	<0.3	tonnes/cap/y	0.44	0.37	10
11	Climate change resilience	>3	Ratio	2.1	1.0	10
12	Renewable energy policy	5	/5	3.7	2.2	10

hh: household, cap: capita, HIC: high-income countries, LIC: low-income countries

In addition, the Green Deal, introduced after the COVID-19 outbreak in 2020, aims to decrease net GHG emissions by at least 55% by 2030 compared to the 1990 level and suggested making changes in the policies of climate, energy, transport, and taxation. It emphasizes effectively implementing a green economy, decarbonization technology, and carbon neutrality. The Singapore Green Plan 2030 advances the national agenda on sustainable development with five key pillars: a city in nature, sustainable living, energy reset, green economy, and a resilient future under a

	People	• Health, Safety, Access to service, Education, Diversity & social cohesion, Quality of housing & built environment
	Planet	• Energy & mitigation, Materials water & land, Climate resilience, Pollution & waste, Ecosystem
	Prosperity	• Employment, Equity, Green economy, Economic performance, Innovation, Attractiveness & competitiveness
	Governance	• Organization, Community involvement, multi-level governance
	Propagation	• Scalability, Replicability

Figure 1.10   Five indicators for sustainable smart cities. Reproduced with permission.[7]

green plan. The Green Deal of India also focuses on five, albeit different, pillars: energy; mobility; industry; green buildings, infrastructure, and cities; and agriculture. These include a fast-tracked approach to green finance and a plan for climate adaptation.

References

1.   http://www.un-documents.net/ocf-02.htm
2.   Jose, R. and Ramakrishna, S. (2021) Comprehensiveness in the research on sustainability, Materials Circular Economy, 3, 1.
3.   https://www.climate.gov/news-features/understanding-climate/climate-change-global-temperature
4.   https://unfccc.int/climate-neutral-now
5.   Stahel, W. R. (2019) *The Circular Economy — A User's Guide*, Routledge.
6.   https://www.corporateknights.com/issues/2023-04-spring-issue/sustainable-cities-index-2023/
7.   Bosch, P., Jongeneel, S., Rovers, V., Neumann, H.-M., Airaksinen, M. and Huovila, A. (2017) CITYkeys indicators for smart city projects and smart cities, doi: 10.13140/RG.2.2.17148.23686.

CHAPTER TWO

SUSTAINABILITY: CURRENT PRACTICES AND EMERGING FIELDS

2.1 Introduction

Sustainable development, as defined in Chapter 1, is comprised of three modules: social, economic, and environmental. These modules address a wide range of current and pressing issues such as pollution, ethical standards, social interactions, health and safety concerns, biodiversity loss, population growth, climate change, and so on. All these aspects of sustainability are intertwined and are now addressed to varying degrees in the field of engineering education and research.

SUSTAINABLE DEVELOPMENT

"Sustainable development is development that meets the needs of the present, without compromising the ability of future generations to meet their own needs."
— *UNSDG*

BIOCAPACITY

The ability of a biologically active region to create a constant supply of renewable resources and absorb excess pollutants is referred to as biocapacity.

This chapter starts by discussing the fundamentals of unsustainable activities, with an emphasis on the environmental, economic, and social implications. Unsustainable practices regarding significant concerns impeding human progress toward the United Nations' definition of sustainable development are examined. The chapter then digs further into the sustainability components of engineering and scientific research, and discusses the academic gaps that exist between sustainable development and engineering. Finally, it introduces six distinct emerging fields that need immediate attention and discusses their associated difficulties and prospects.

2.2 Unsustainable Economic Practices

Economists have used economic indicators to forecast future economic growth and assess a country's overall economic performance by analyzing gross domestic product (GDP) to connect GDP and sustainability. Global GDP continues to expand, as seen in Figure 2.1, and reached 84.5 trillion dollars in 2020.[1] On the other hand, the world's biocapacity is dwindling to 1.5 hectares per person.[2]

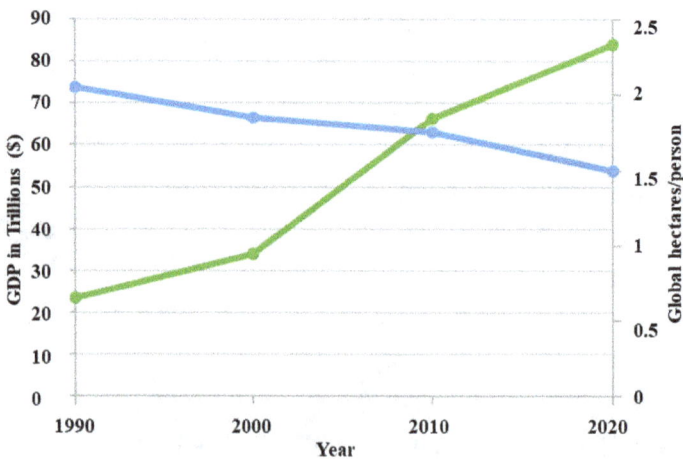

Figure 2.1   Global GDP vs economic sustainability. Reproduced with permission.[1]

Since the post-World War II period, our economy (global GDP) has been rising and major technological developments have occurred since the 1960s. Scholars have investigated a link between GDP and biocapacity and demonstrated how, over the last century, our economic activity has had a substantial influence on our global biocapacity because of many new technological breakthroughs.[3,4] When GDP is compared to biocapacity, global GDP growth is shown to be unsustainable and ignores the environmental dimensions of sustainable development.

It might be concluded that the decline in global biocapacity is a direct result of human activities, such as the growing need for urban space and food. Because of the current state of our economy, global consumption requires more than 1.5 planet Earth to support itself, and this will rise as the middle class grows.[5] Also already being felt are the negative environmental externalities associated with present living standards in developed countries, such as climate change and rising sea levels as a result of a high rate of natural resource exploitation.

2.3 Unsustainable Social Practices

The social dimension of sustainability has often been overlooked, with a disproportionate focus placed on economic growth, particularly the expansion of GDP and national wealth. As a result, social well-being has been neglected, leading to serious and widespread consequences. These include social injustice, civil unrest, inequality, unemployment, gender disparity, inadequate housing, and health insecurity — all of which undermine the foundation of a sustainable society. To put this in context, in 2021, over 9% of the worldwide workforce from economically deprived families lived in extreme poverty (Figure 2.2) and were unable to maintain a stable source of income.[6] Globally, the loss of human capital resources as a result of gender imbalance was projected to be $160 trillion in 2018, according to the World Bank.[7]

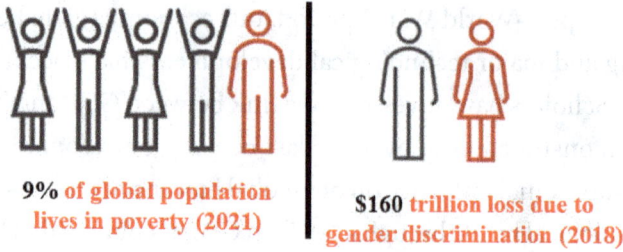

9% of global population
lives in poverty (2021)

$160 trillion loss due to
gender discrimination (2018)

Figure 2.2   Unsustainable social practices. Reproduced with permission.[7]

The fundamental purpose of social sustainability is to achieve societal goals and good working circumstances, and it is considered that by leveraging new developing areas of engineering and science, we can undoubtedly solve a few conundrums of social sustainability. For instance, we are on the verge of fully depleting the planet's fossil fuel reserves. If the earth proceeds toward a greener future, six million people would lose employment as a consequence of the fall in the use of fossil fuels. To properly balance this, the International Labor Organization proposes that a green economy, particularly new technical advances, may provide 24 million new employment opportunities by 2030 if sustainable development policies are matched.[8]

2.4 Unsustainable Environmental Practices

The linear economy model was introduced to quantify human impact on the environment, as represented in Figure 2.3. It suggests that humans have historically followed a take-make-dispose approach, ignorant of the notion of recycling or the adverse effects on the environment, planet, or humans. As a result, significant difficulties are emerging on the horizon because of the negative consequences of *resource depletion, environmental pollution, and exponential population growth.*

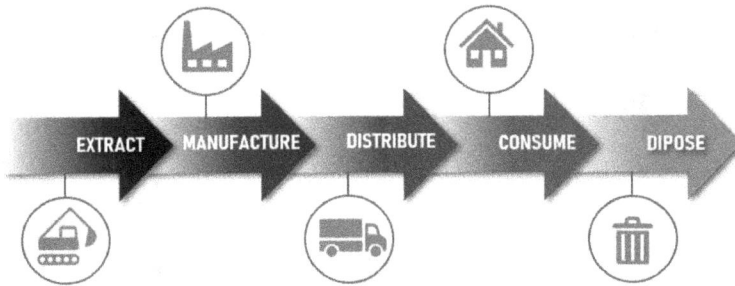

Figure 2.3   Linear economy model. Reproduced with permission.[9]

Consider resource depletion as an example; our economy is mainly fueled by coal and oil, which are natural resources that are difficult to synthesize. Conservatively, according to estimates and proven resources, our coal reserves would presumably last us until 2150, and oil reserves would sustain us probably until 2060 to 2070.[9] On the other hand, human activities since the Industrial Revolution and fast modernization have had a substantial impact on the planet, accelerating environmental deterioration. For instance, about 165 million tonnes of plastic are believed to be circulating in the marine environment.[9] Additionally, according to a recent United Nations survey, the world's population is around 7.8 billion in 2020 and is expected to expand by approximately one billion by 2025 and surpass 9.7 billion by 2050.[10] Hence, it would be necessary for the planet to increase food production by 70%. This expanding need for food would impose a significant strain on the environment, resulting in increased CO_2 emissions, deforestation, and the overuse of freshwater. As seen in Figure 2.4, the increasing concern is global warming; because of rapid industrial expansion and the emission of greenhouse gases, considerable changes in our climate have occurred. According to the World Meteorological Organization, Earth has been quickly warming since the 1970s, signaling that human activities are harming the planet and environmental awareness is being neglected as a result of unsustainable environmental practices.

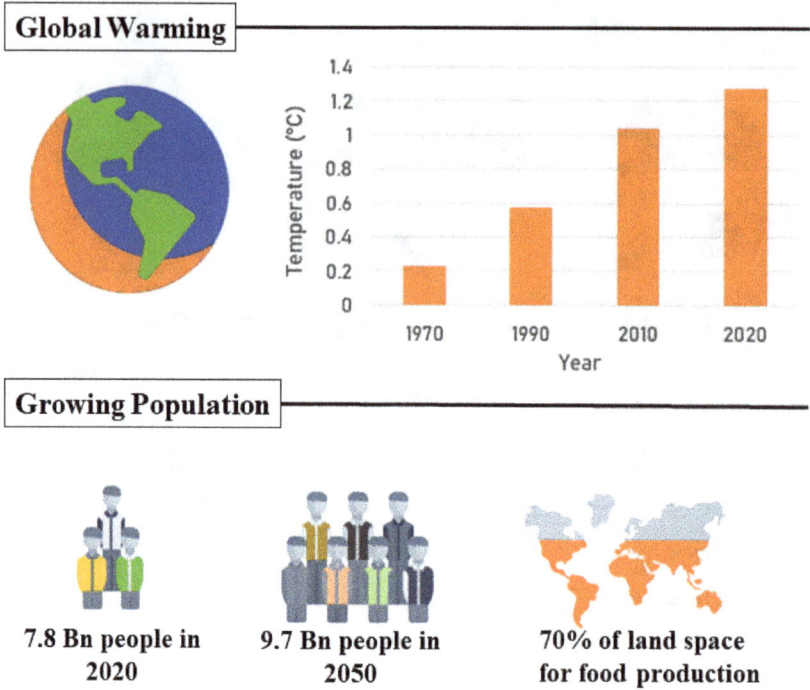

Figure 2.4   Effects of rising population and global warming. Reproduced with permission.[10]

2.5 Sustainability in Academic Practice

The sustainability challenge is one of the difficulties that engineering education faces in the 21st century. In other words, there is a fundamental lack of integration of sustainability principles that promote Economic, Social, and Environmental (E, S, E) facets. As the Bruntland study rightly stated in 1987, sustainability is defined as *"development that meets the needs of the present without compromising the ability of future generations to meet their own needs"*.[11] In this spirit, the engineering curriculum is accountable for preparing engineers to address both present and future challenges outlined earlier in this chapter. The goal is to characterize the issue, concentrating on all dimensions of sustainability (E, S, E), and to transform the challenges into an accessible format by applying engineering principles, values, knowledge, and competencies to everyday operations to discover appropriate solutions.

Undoubtedly, significant progress has been achieved in recent years toward creating a more environmentally conscious curriculum in engineering institutions. Owing to the unconnected dots of tools, regulations, and practices, there is still a lack of a comprehensive approach to understanding sustainability concepts in academia.[12] We must properly connect the dots and provide the opportunity for students to study, evaluate, make knowledgeable choices, and apply what they have learned to real-world circumstances. On the other hand, problem-based learning is still lacking to inspire engineers to understand the context of sustainability.

2.5.1 Sustainability Policies and Principles

As shown in Figure 2.5, four main pillars need to be integrated into the sustainability curriculum to equip future engineers and scientists with industrial-relevant skills and competencies. *Sustainable Policies and Principles* are fundamental concepts for students to grasp in academia

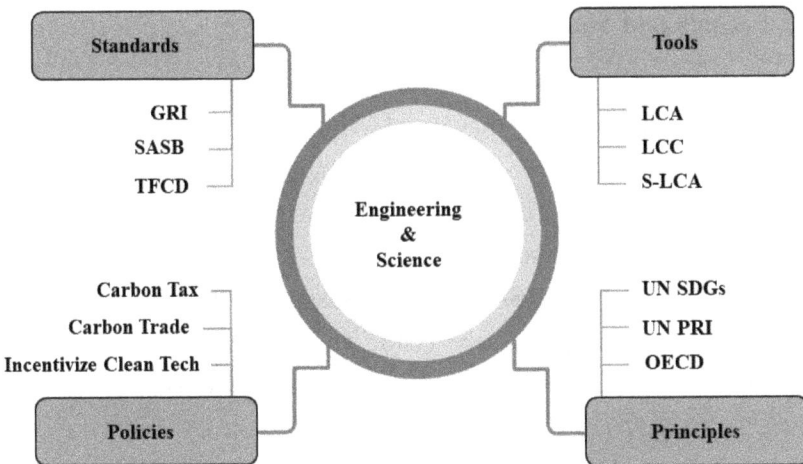

Figure 2.5 Engineering and science education focusing on sustainability aspects. See main text for explanation of abbreviations.

before embarking on industrial settings to tackle real-time challenges. Policies vary by country, and hence a basic understanding of regional policies such as carbon taxation, carbon trading, and promoting clean technology is required. Among all policies, carbon taxes have been extensively implemented in many nations. For instance, the United Kingdom has implemented an annual carbon tax on high-emitting fossil fuel-powered automobiles. As vehicle age increases, the carbon price increases, incentivizing users to convert to clean energy cars.

Along with policies, the United Nations' Sustainable Principles are essential to grasp at the academic level. It is usually a good way to start with the 17 United Nations Sustainable Development Goals (UN SDGs) listed in Chapter 1 since this provides a foundation for understanding global sustainability activities. Additionally, the United Nations established six foundational rules for sustainable investment via the United Nations Principles for Responsible Investment (UN PRI) to guide future prospective investment in sustainable projects.[13] Fundamental knowledge is always necessary to comprehend the context of sustainability at the policies and principles level.

2.5.2 Tools and Standards

The most crucial component of the sustainability curriculum to examine is the tools and standards; they provide future engineers and scientists with the requisite skill sets demanded by the industry. Thus, we will aim to gain a comprehensive understanding of each of them. To begin, tools such as Life Cycle Assessment (LCA), Life Cycle Costing (LCC), Social Life Cycle Assessment (S-LCA), and Sustainability Reporting using the Global Reporting Initiative (GRI) and Sustainability Accounting Standards Board (SASB) are required to meet the UN SDGs.

2.5.2.1 Life Cycle Assessment

Consider the plastic waste scenario outlined above; 165 million tons of plastic waste are now distributed in the marine biosphere. While we recognize that plastic carry bags have a short useful life, the Stop Plastics group estimates that each kilogram of plastic emits six kilograms of CO_2.[14] This level of technical assessment and parametric computation is only achievable via an analysis of the plastic waste life cycle scenario using commercial software such as Simapro or Gabi, or open-source tools such as OpenLCA. As environmental awareness grows internationally, engineers must practice using these methods in product development at the design level to assess environmental impact.

2.5.2.2 *Life Cycle Costing*

The term "life cycle costing" refers to the process of determining the total cost of ownership of an asset across its useful life from cradle to grave. For example, owning a vehicle has a lifelong cost in addition to the original cost, such as fuel, maintenance, and parking. Notably, the building industry is responsible for over 39% of worldwide CO_2 emissions, or about 16.7 billion tons.[15] Developed nations have used LCC analysis to acquire green building certification, effectively leveraging circular economy ideas such as disintegration, refurbishment, and reuse concepts. However, this kind of information combined with real-world examples, particularly when applying tools like LCC assessment across all engineering and scientific fields to meet UN SDGs, is inadequate. Hence, we may have to introduce LCC analysis in every aspect of engineering and science projects to estimate the financial risk and environmental benefits.

2.5.2.3 *Social Life Cycle Assessment*

The S-LCA method is used to assess a product's social and sociological characteristics. This technique, which is similar to the environmental LCA, aids us in understanding how products or projects benefit or impact social circumstances such as employees, local communities, customers, and stakeholders. It is rather challenging to operationalize S-LCA. This is mainly because the approach to environmental (LCA) and economic (LCC) evaluation evolved in quantitatively focused scientific and technological fields, but social problems are difficult or impossible to measure owing to the complex nature of business and human interactions. As UN Sustainable Development Objective 3 highlights the goal of health and well-being, we might utilize and appropriately use this principle while developing new products or initiatives.[16]

2.5.3 Standards

Along with financial and non-financial reporting, firms are increasingly reporting on sustainability, often known as Corporate Social Responsibility or Environmental, Social, and Governance (ESG) reporting. The importance of sustainability reporting is that it guarantees that businesses assess their influence on environmental concerns and allows them to be transparent about the risks and opportunities they face. Around 90% of global large corporations are currently disclosing their ESG measures. The practice is progressively catching on among small- and medium-sized enterprises (SMEs) who are following in the footsteps of bigger corporations. The benefit of ESG reporting is that it allows businesses to demonstrate their transparency while also solving global challenges by lowering social and environmental consequences. As illustrated in Figure 2.5, widely recognized standards such as the GRI, the SASB, and the Task Force on Climate-related Financial Disclosures (TCFD) provide a comprehensive, adaptable framework for businesses of any size to report on their economic, environmental, and social impacts. Additionally, the International Organization for Standardization (ISO) provides standards and certification to achieve sustainable development, also known as sustainability standards as shown in Table 2.1.

2.6 Emerging Fields Focusing on Sustainability

Sustainable development has become a concern for governments since humanity's influence on the environment has escalated significantly over

Table 2.1   ISO standards used for SDGs and sustainability elements.[17]

ISO Standard	SDG	Sustainability Element
ISO/Guide 82:2014: Addressing sustainability in standards. ISO 26000: Social responsibility guidance. ISO 20400: Sustainability procurement guidance. ISO 20121: Event sustainability management systems.	1–16	**1. Business Sustainability** ISO 9001; ISO 31000; ISO 31010; ISO Guide 73.
ISO 37120: Indicators for city/community services and quality life.	1	**2. Social Sustainability** ISO 26000; Occupational
ISO 22000: Food safety management. ISO 4002: Equipment for sowing and planting. ISO 4197: Equipment for working the soil. ISO 6880: Agriculture machinery. ISO 20635: Infant formula and adult nutrition. ISO 1871: Food and feed product.	2	Health and Safety Management System 18001.
ISO 11137: Sterilization of healthcare products. ISO 8828: Implants for surgery. ISO 18615: Traditional chinese medicine. ISO 11073: Health informatics.	3	**3. Environmental Sustainability** ISO 14001.
ISO 18091-2014: Quality management system (application of ISO 9001: 2008 in local government). ISO 11228-2-2007: Ergonomics.	5	
ISO 14046: Water footprint. ISO 15839: Water quality. ISO 24516-1: Drinking water distribution networks. ISO 24518: Crisis management of water utilities. ISO 24526: Water efficiency management systems.	6	
ISO 50001: Energy management. ISO 9553: Solar energy. TR 10217: Solar water heating system. ISO 13065: Sustainability criteria for bioenergy. ISO 17743: Energy savings. ISO 20619: Calculation methods for energy savings.	7	
ISO 45001: Health and safety. ISO 37001: Anti-bribery management systems.	8	
ISO 37154: Smart community infrastructures and transportation. ISO 21542: Building constructions. ISO 50501: Innovation management system.	9, 10	

(Continued)

Table 2.1　(*Continued*)

ISO Standard	SDG	Sustainability Element
ISO 37101: Sustainable development of community (management system).	11	
ISO 28000: Security management.		
ISO 37104: Strategies for smart cities and communities.		
ISO 37120: Sustainable development of community (indicators for city services and quality life).		
ISO 10711: Intelligent transport system.		
ISO 12720: Sustainability in building and civil engineering works.		
ISO 14001: Environmental management system.	13	
ISO 14044: Life cycle assessment.		
ISO 14064: Greenhouse gases.		
ISO 14080: Frameworks for methodologies on climate actions.		
ISO 10987: Sustainability on earth-moving machinery.		
ISO 30000: Ship recycling.	14	
ISO 29400: Sips and marine technology (offshore wind energy).		
ISO 21070: Marine environment protection.		
ISO 35101: Petroleum and natural gas industries (Arctic operations).		
ISO 12878: Environmental monitoring of the impacts from marine finfish farms.		
ISO 19900: Offshore structure requirements.		
ISO 14055: Combating land degradation and desertification.	15	
ISO 38200: Chain of custody of wood and wood-based products.		
ISO 23611: Soil quality (sampling of soil invertebrates).		
ISO 15592: Effects of pollutants on juvenile land snails.		
ISO 22030: Soil quality (biological methods).		
ISO 11850: Machinery for forestry (safety requirements).		
ISO 19600: Compliance management system.	16	

the last century because of a rapidly growing population and concurrent steep declines in natural resources. **Pollution and waste management**, in addition to climate change, energy crises, and resource shortages, are the monumental issues that mankind will confront in the coming years. When we dissect these two major issues of pollution and waste management into their fundamental components, we discover the following: *plastic pollution, water management, electronic waste, municipal trash, food waste, and construction & demolition waste* all contribute, and we as researchers have the opportunity to address the challenges, as depicted in Figure 2.6 and outlined in the following chapters.

2.6.1 *Sustainable Electronic Waste (E-Waste) Management*

Sustainable e-waste management comprises formal collection, treatment, and recycling to recover critical and precious metals. This will reduce waste disposal and recover precious metals from e-waste. However, the informal sector recycles 90% of e-waste, creating unfavorable circumstances. Informal recycling procedures are harmful to the environment and human health. Nevertheless, as shown in Figure 2.6, novel recycling technologies based on advanced techniques are gaining traction across the world, and the circular economy idea is incorporated

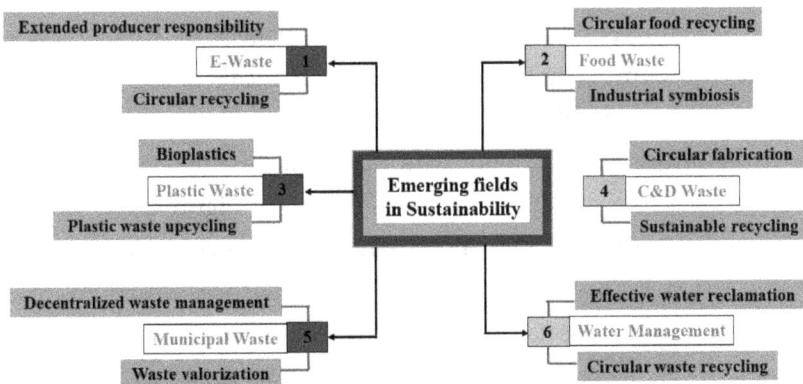

Figure 2.6 Emerging fields in sustainability and opportunities.

into e-waste management. Furthermore, in many countries, Extended Producer Responsibility is being legislated, which implies that electronic makers are responsible for collecting and sustainably recycling electronic trash at the end of its useful life. In the next chapter, novel conventional approaches and possibilities for sustainable e-waste management will be examined in depth.

2.6.2 Sustainable Food Waste and Loss Management

Waste generation in the food supply chain has become a serious global concern in the 21st century. Due to population growth and rising scarcity of natural resources, this results in a rise in not just economic costs, but also social costs and a detrimental effect on the environment. Researchers are tackling food waste in several ways; topics include reducing food waste via circular principles, most notably the industrial symbiosis approach, anaerobic digestion, and compositing to minimize landfills. To achieve long-term sustainable food waste and loss management, a comprehensive framework for developing an efficient and sustainable food supply chain is essential. Chapter 4 will explain the LCA and circular concepts that govern our present food waste and loss management.

2.6.3 Sustainable Plastic Waste Management

Plastic recycling is critical in its management because if not recycled on time, it becomes combined with other chemicals or materials, making recycling more difficult and becoming a source of pollution. There are two approaches to dealing with plastic waste and promoting sustainable plastic waste management, as shown in Figure 2.6: (i) Plastic waste upcycling entails four distinct recycling procedures, as defined by the ASTM D7209 standards, and (ii) bioplastics, which are made from plant-based materials and are used to replace non-biodegradable polymers. These topics will be discussed in further depth in Chapter 5 on plastic waste management.

2.6.4 Sustainable Construction and Demolition Waste Management

Although construction and demolition (C&D) waste management is common, there are always areas for improvement. For example, emerging research fields such as circular manufacturing, which focuses on modular fabrication and installation, allow for easy disassembly and disintegration at the end of life. Over 90% of C&D waste is recycled to recover valuable resources such as ferrous metals, non-metallic minerals, plastics, and wood. However, developing countries often face insufficient management of building and demolition waste or repurposed materials, resulting in the majority of waste ending up in landfills. In this study path, new initiatives examining existing difficulties and potential with a strong emphasis on sustainability are explored in depth in Chapter 6 on C&D waste management.

2.6.5 Sustainable Municipal Solid Waste Management

As the planet becomes more urbanized, problems and expectations associated with meeting incremental infrastructure requirements and services such as waste collection and disposal naturally increase, particularly in urban regions. This necessitates a paradigm change away from the linear economy and toward a circular economy, which mimics nature by repurposing waste. By leveraging circular economy concepts and waste valorization, we may encourage consumption reduction by considering a product's useful life and, ultimately, its reuse or recycling. Chapter 7 on sustainable municipal solid waste (MSW) management discusses how this notion of sustainability and the circular economy has advanced the MSW management system beyond simple disposal to recycling and resource recovery.

2.6.6 *Sustainable Water Management*

Water is a necessary element of all life forms. According to estimates, 1.8 billion people would live in nations with absolute water shortage by 2025, and two-thirds of the world's population will live in water-stressed areas. The demand for more water from existing natural surface and groundwater sources has risen due to increased population and a better quality of life in communities. As a result, comprehensive water recycling and reclamation are required. The water management chapter (Chapter 8) offers appropriate sustainable practices, focusing on water pollution, water shortages, and wastewater management.

References

1. https://www.statista.com/statistics/268750/global-gross-domestic-product-gdp/.

2. https://www.overshootday.org/content/uploads/2021/06/Earth-Overshoot-Day-2021-Nowcast-Report.pdf.

3. Kubiszewski, I. et al. (2013) Beyond GDP: Measuring and achieving global genuine progress, Ecological Economics, 93, 57–68.

4. Gabbi, G. et al. (2021) The biocapacity adjusted economic growth. Developing a new indicator, Ecological Indicators, 122, 107318.

5. https://think.ing.com/uploads/reports/Financing_the_Circular_Economy.pdf.

6. https://devinit.org/documents/1343/Economic_poverty_factsheet_June_2023.pdf.

7. Wodon, Q. and de la Briere, B. (2018) Unrealized Potential: The High Cost of Gender Inequality in Earnings. The Cost of Gender Inequality, Washington, DC: The World Bank.

8. https://www.un.org/sustainabledevelopment/blog/2019/04/green-economy-could-create-24-million-new-jobs/.

9. Selvan, R. T. and Ramakrishna, S. (2022) *Sustainability for Beginners: Introduction and Business Prospects*, World Scientific.

10. https://www.un.org/en/desa/world-population-projected-reach-98-billion-2050-and-112-billion-2100.

11. Keeble, B. R. (1988) The Brundtland report: 'our common future,' Medicine and War, 4(1), 17–25.

12. Guerra, A. (2017) Integration of sustainability in engineering education: Why is PBL an answer?, International Journal of Sustainability in Higher Education, 18(3), 436–454.

13. https://www.unpri.org/download?ac=5330.

14. https://stopplastics.ca/wp/resources/carbon-footprint-of-plastic/.

15. Ali, K. A., Ahmad, M. I. and Yusup, Y. (2020) Issues, impacts, and mitigations of carbon dioxide emissions in the building sector, Sustainability, 12(18), 7427.

16. Klöpffer, W. (2005) Life cycle assessment as part of sustainability assessment for chemicals, Environmental Science and Pollution Research, 12(3), 173–177.

17. https://iso26000.info/sustainability-standards-from-iso/.

CHAPTER THREE
MATERIAL SUSTAINABILITY IN ELECTRONICS AND ELECTRICAL MOBILITY

3.1 Introduction

Electrical and electronic devices have become increasingly prevalent in modern life, enhancing convenience and serving a wide range of daily needs. Since the early 2000s, demand for these devices has grown significantly, driven by several factors, including declining product costs, shorter product life cycles, and increased purchasing power of lower-middle and middle-income group countries. This has triggered a sizable growth in the global electrical and electronic market, which is expected to expand at a rate of 6% in 2021. The electrical and electronics industry comprises a variety of products such as light-emitting diodes (LEDs), consumer electronics, electrical household appliances, electronic medical equipment, microelectronic components, software, on-board diagnostics, in-car touch screens, cameras, automotive vehicles, and navigation systems.

Additionally, in recent years, electric mobility (e-mobility) has emerged as a powerful and rapidly growing sector, focused on the efficient use of energy for transportation. It aims to address major global challenges such as climate change, pollution, and structural transformations in the energy sector's supply chains. In support of a low-carbon economy, there is a growing emphasis on utilizing renewable energy sources — such as solar and wind power — to limit the rise in global temperatures to within 1.5°C. These renewable systems rely heavily on electrical and electronic technologies to operate effectively. For instance, the global demand for electric vehicles (EVs) reached 10 million in 2022 and is expected to grow 21 million units by end of 2025. The EVs currently available, such as battery EVs, plug-in hybrid EVs, hybrid EVs, and fuel cell EVs, minimize overall greenhouse gas emissions and decarbonize on-road transportation.[1] In order to supply the energy to run EVs, rechargeable batteries made of lithium iron phosphate, lithium nickel-manganese-cobalt oxide (NMC), lithium titanate oxide, lithium nickel-cobalt-aluminum oxide, nickel-metal hydride, and lead-acid materials are required.

The rapid transition towards the usage of modern advanced technologies creates immense pressure on the raw materials required in the manufacturing of new electrical and electronic devices including rechargeable batteries. The situation becomes more alarming as the demand for these critical raw materials for clean energy production and advanced technology application is expected to grow continuously. Consequently, their prices fluctuate a lot on a daily basis which make the manufacturing costs unpredictable. It is projected that the total material requirement will increase by around 200–900% in the electricity sector and 350–700% in the transport sector from 2015 to 2050 (Figure 3.1). The growing demand and limited supply of resources pose serious challenges for both urban and industrial sectors, highlighting the urgent need for suitable and sustainable solutions. Developing effective strategies for the sustainable use of electrical and electronic devices as energy resources requires a comprehensive evaluation of the three key pillars of sustainability: society, economy, and the environment.

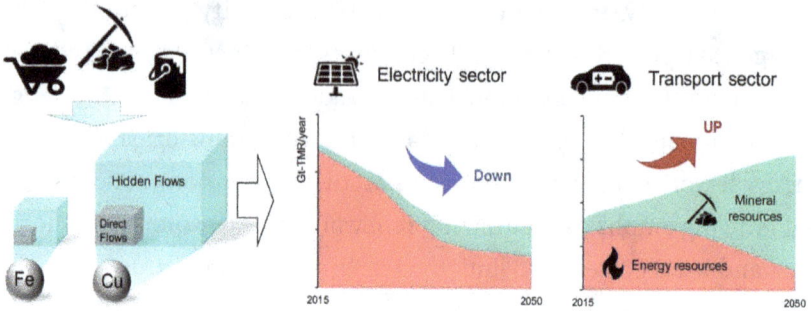

Figure 3.1   Total material requirement increases from year 2015–2050.

3.2 Critical Metals in the Electronic Mobility Sector

Metals have played a vital role in the advancement of human civilization. From ancient times to the modern digital era, the influence of metals on societal progress is clearly visible. Scientific developments — including the discovery of new elements, the synthesis of advanced materials, and their application in emerging technologies — have evolved in parallel with time. From the landmark discovery of fire as an energy source to the development of rechargeable batteries and renewable energy devices, the ways in which energy is produced, stored, and distributed have undergone significant transformation. Metals and metal compounds have been instrumental in enabling these advancements. Figure 3.2 illustrates the evolving role of metals in energy generation and, subsequently, in e-mobility.

As the use of metals in electrical and electronic devices, including e-mobility applications, has grown significantly over time, their demand in the global commodity market has surged — despite their limited supply. This imbalance has created a widening gap between supply and demand, resulting in growing pressure on the energy sector and increasing the "criticality" of certain metals, which can be defined as follows[17]:

$$\text{Criticality} = \text{Supply risk} \times \text{Vulnerability}$$
$$= \text{Supply disruption} \times \text{Economic consequences} \quad (3.1)$$

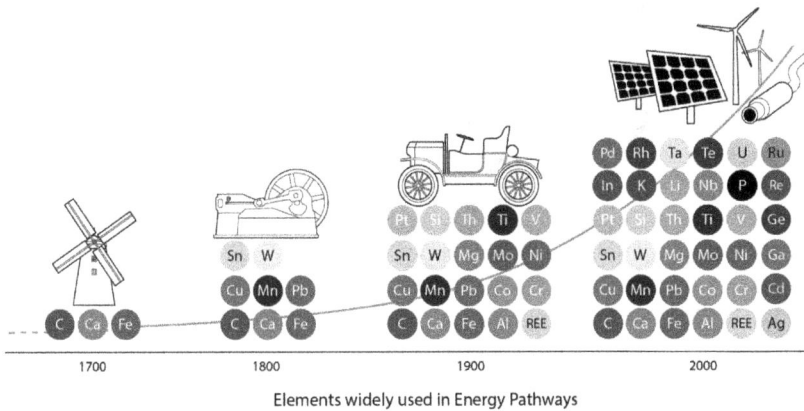

Figure 3.2 Evolution of the role of metals in energy generation and e-mobility. Reproduced with permission.[16]

Due to the essential role of metals in the clean and green energy sector, particularly in e-mobility, criticality assessments are carried out to evaluate the potential supply risks and market availability within specific countries or regions. While criticality may vary between nations, the lists published by the European Union (EU) and the U.S. Department of Energy are widely recognized.

Figure 3.3 highlights the supply concerns for critical metals from an EU perspective. It shows that approximately 20 countries control the supply of around 90% of the world's most important metals, placing the EU at significant supply risk for these materials. At the same time, global demand for metals and minerals is growing at a rate exceeding 8% per year. Therefore, such metals are classified as "critical" for the EU.

As a result, the intensified consumption of critical metals demands the opening of new mines and refineries to meet future needs. However, this is a costly and resource-intensive process. Moreover, mining and refining activities can cause severe environmental and ecological damage, rendering the system unsustainable due to its inherent vulnerabilities.

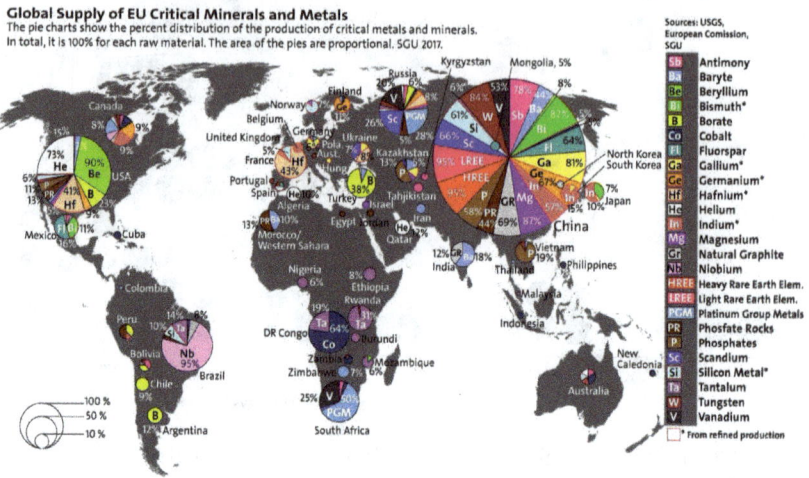

Figure 3.3   The global supply of critical minerals and metals. Adapted with permission.[2]

Copper: In order to understand the burgeoning issue of materials sustainability, copper, a very common metal applied in most electrical and electronic devices, shall be used as an example. An electric motor of a vehicle itself needs 6 km of copper wires to connect with the battery cells.[3] On a weight basis, an EV contains ~80 kg copper which is twice that of a hybrid car (40 kg copper), while a petrol-powered car requires about 20 kg of copper. As estimated by the Copper Development Association, a rise of up to 50% in copper demand is expected in the next 20 years,[4] which can be 10-fold by 2050 if the world shifts towards a low-carbon energy future. By looking at the estimated global copper reserves of 830 million tons (Mt) and its annual demand of 28 Mt,[5] only 30 years are left. However, adding the present recycling rate of copper (i.e., 35% of the copper) can significantly reduce the dependency on mined copper, as copper is inherently a circular material without any loss of quality when reused for another function. The progress in recycling more copper will not only help meet demand but also move towards achieving material sustainability and conserving natural resources.

Cobalt: The sustainability of cobalt is the biggest concern for EV manufacturers as the present market of lithium-ion batteries (LIBs) is driven by this metal. LIBs consume approximately 62% of worldwide cobalt production in 2020 as the integral part of different cathode materials, which was only 20% in 2006, crossing over 50% in the year 2016.[3] A total demand for cobalt has been found to be ~125,000 tons in 2016, which is forecasted to double in the next decade with over 120,000 tons of cobalt alone needed in rechargeable batteries (Figure 3.4). On the other hand, it is predicted that ~80% of cobalt supply worldwide will originate in the Democratic Republic of Congo (DRC). Despite the military conflict and involvement of child laborers in artisanal mines, the world will continue to be dependent on DRC for cobalt production and supply. As per the report presented by Green Car Congress, the fulfillment of future cobalt supply will depend upon the diversified mining of its companion metal reserves like nickel, copper, and platinum-group metals (unfortunately all of them are individually designated as critical metals for the sustainable developments of modern low-carbon technologies). That is why the economic value of NMC cathode-containing batteries ($1684/ton) is higher than those of $LiNiO_2$ cathode ($523/ton) and $LiMn_2O_4$ cathode ($636/ton).

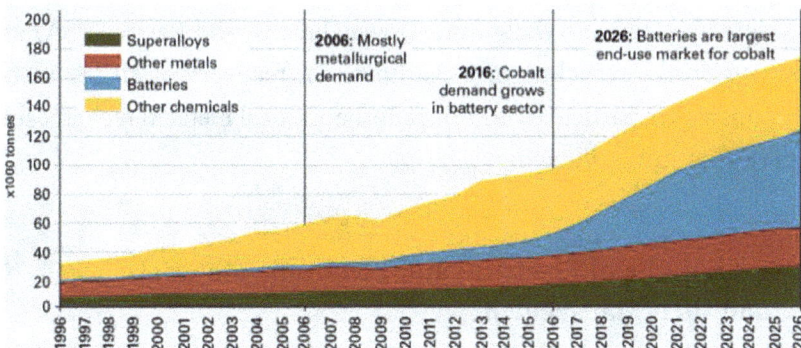

Figure 3.4   Cobalt demands for its usage in different industrial sectors. Adapted with permission.[3]

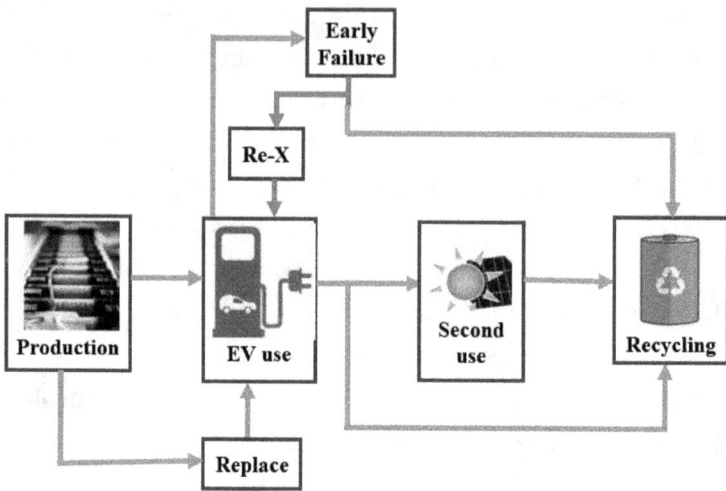

Re-X: repair, reuse, or remanufacture

Figure 3.5 Typical life cycle stages for the recycling of most in-use LIBs.

Therefore, recycling spent batteries by assessing their life cycle (as depicted in Figure 3.5) is getting more attention due to the associated materials' economic values. Recycling policies in several countries are being implemented more specifically to optimize resource recovery and material management of end-of-life (EoL) EV batteries. A few major policies in this regard have been summarized in Table 3.1. Some leading recyclers like Umicore (Belgium), Dowa (Japan), and Retriv (USA) are working on the recycling of LIBs; however, recovery of all the metal components (in particular, lithium, another critical metal in rechargeable batteries) is still challenging.

3.3 End of Life of Electrical and Electronic Materials and their Management

After completing the EoL of all the electrical and electronic equipment, what is discarded/broken/surplus/obsolete is known as e-waste or waste from electrical and electronic equipment (WEEE). The European Commission has defined WEEE as wide-ranging EoL commodities

Table 3.1   Policies and events on recycling EV batteries by different countries.

	Country	Policies and Events	References
1	China	Short-term measures were taken for the management and recycling of batteries and utilization of power batteries in new EVs (including design, production, and recycling responsibilities).	Ma et al., 2018
2	European Union	Extended producer responsibility (EPR) system is implemented for the collection and recycling of batteries through directives.	European Commission, 2019
3	Germany	German Batteries Act emphasizes to collect EoL batteries by all manufacturers, importers, and accumulators of batteries.	Umweltbundesamt, 2019
4	Japan	Voluntary action is performed by the manufacturer to collect and recycle batteries under the law for the promotion of effective utilization of resources. In 2004, a producer responsibility organization was established, known as the Japan Battery Recycling Centre, which recycles waste batteries. However, there is no specific law for battery recycling.	Winslow et al., 2018
5	India	The EPR rule and E-waste (Management) Amendment Rules, (2018) have provided different categories to recycle e-waste including batteries. However, there is no specific law for recycling EV batteries.	7
6	Republic of Korea	Consumers are getting a subsidy to return EoL batteries for recycling. However, there is no specific law for battery recycling.	8
7	Singapore	Two recycling plants have started recycling LIBs for EVs. However, there is no specific law for battery recycling.	9
8	USA	The universal waste rule was implemented to collect a large number of batteries. There is no federal law for recycling of EoL EV batteries. A few states ban the disposal of batteries into landfills.	Winslow et al., 2018

employed to produce, measure, and transfer electrical or electromagnetic current during their service life.[10] On the other hand, e-waste definition is given by The Basel Action Network and Silicon Valley Toxics Coalition as "a broad and growing range of electronic devices ranging from large household items to personal computers which have been discarded after their EoL".[11] The Association of Plastics Manufacturers in Europe gives one more definition: "e-waste is the multiferrous combination of ferrous, non-ferrous, ceramic, and plastic materials". Seven categories of e-waste based on their application are shown in Table 3.2.

Table 3.2   Different categories of e-waste as per their applications.

Heat and temperature exchanger equipment including refrigerators, air conditioners, and heat pumps	
Display screen and monitors including computers, laptops, televisions, notebooks, and tablets	
Electric lamps including fluorescent, high intensity discharge, and LED lamps	
Large equipment including electric stoves, washing machines, Xerox machines, and photovoltaic panels	
Small equipment mostly including household items such as microwaves, toasters, vacuum cleaners, electric kettles, calculators, video cameras, electrical toys, and small medical devices	
Small information and communications technology equipment including mobile phones, routers, table printers, global positioning systems, personal computers	
Electric vehicles, energy storage devices	

With an annual growth rate of more than 4%, e-waste is enormously growing in volume around the globe. In 2019, globally ~53.6 Mt of e-waste was generated, which represents an increase of 9.2 Mt since 2014 level (equivalent to 7.3 kg increase per inhabitant), and is estimated to reach 74.7 Mt by the end of year 2030.[12] At the regional level, Asia is the top generator of e-waste (~46%), followed by the USA (~24%), and Europe (~22%); however, Europe is top per inhabitant. The e-waste generation gap is very wide between the developed and developing countries. The richest nation is generating 28.5 kg/inhabitant contrary to the poorest nation of only 2.5 kg/inhabitant. Ironically, developed nations have shielded themselves by exporting their e-waste to developing and under-developed nations, which are still struggling to build effective e-waste management systems. In most developing nations (specifically low- and middle-income nations), a large percentage of e-waste is either disposed of in landfill sites or handled by the informal recycling sector. The burning of waste wires and printed circuit boards (PCBs) is a common practice to remove the metals and polymers. Elementary practices that include hand-picking and dismantling, nitrate/aqua-regia leaching, and throwing the residual and effluent streams in an open environment, are also done to retrieve the precious metals from e-waste. Therefore, safeguards to the environment and public health are lacking along with an inappropriate efficiency of resource recycling.

3.3.1 Global Initiatives for E-Waste Legislation

Globally, electrical and electronic equipment are considered waste once their EoL is completed and discarded from developed to developing nations either as used materials (in the name of charity) or in the form of e-waste. Improper handling of e-waste is common in most developing nations and demonstrates that they do not have proper guidelines for e-waste management through their legislation. As a result, the informal sectors play significant roles in low- and middle-income nations

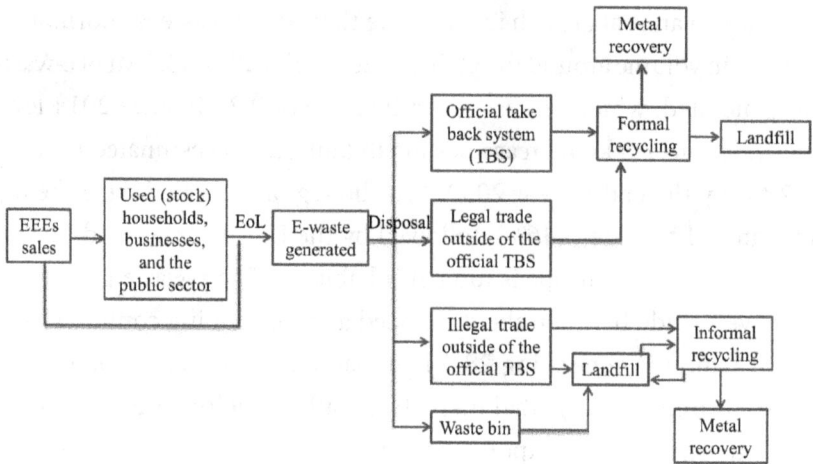

Figure 3.6 Pictorial representation of a typical e-waste management system.

to manage e-waste. Figure 3.6 depicts the practices of a typical e-waste management system (legally and illegally) where it is collected from sales, households, business and public sectors after attaining the EoL. Globally, formal e-waste initiatives have been proposed to manage the e-waste and restrict its transboundary movement. Table 3.3 shows some of the legislation practices developed by the leading countries for sound and sustainable e-waste management within their legislative limit.

3.3.2 Sustainable E-Waste Management

Sustainable management of e-waste includes formal collection, handling, and recycling, a post-consumer operation to reclaim critical and precious metals from e-waste and also lessen the amount of waste disposal. It has been noted that the informal sectors do 90% of e-waste recycling under adverse conditions. The irrational and unsafe informal recycling practices pose a serious threat to the environment as well as human health. Open-air burning, sorting, and dismantling in a small room, and use of toxic chemicals for reclaiming precious metals damage

Table 3.3 Global initiatives for management and legal framework of e-waste.[13]

Country/Region	Legislation/Regulation	Description
South Korea	Act on the control of transboundary movement of hazardous wastes and their disposal, 1994	Restriction on export without consent from the importing country
Basel (Switzerland)	Basal Convention, 2002	To ban hazardous and toxic substances in e-waste through the transboundary moment
Belgium	Directive 2002/96/EC on WEEE, 2002	The Public Waste Agency of Flanders controls the waste management and responsibility of producer
Finland	Government Decree on WEEE, 2004	Export prohibited out of the European Union (EU) unless exporter proves that reuse and/or recycling will be practiced as directed in this Decree
France, Germany and the Netherlands	Under EU directives in 2005	Limited use of toxic materials by producers; collection and processing of used electronics by distributors and municipalities; France introduced an "eco-cost" for treating WEEE
Japan	Law for the control of export, import and others of specified hazardous and other wastes	Export prohibited without consent from the import country
Vietnam	Law on environmental protection, 2005	Prohibits the movement of hazardous waste from abroad; stipulates responsibilities for waste generator
China	Catalogue of restricted imports of solid wastes, 2008	Restriction on junk electrochemical products and electrical wires mainly for copper recycling
Norway	Revised EU directives, 2006	A WEEE register established with mandatory membership for every producer and importer of an approved take-back company
United Kingdom	Under EU directives in 2007	Adopted the EU directives

(Continued)

Table 3.3   (*Continued*)

Country/Region	Legislation/Regulation	Description
Thailand	The criterion for import of used EEE (UEEE), 2007	Control on the classified UEEE
Singapore	Import and export of e-waste and used electronic equipment, 2008	Approval on the movement of hazardous e-waste on case-by-case basis
Pakistan	Import policy order, 2009	Banned the import of refrigerators and air conditioners; cathode ray tubes (CRTs) can be imported only with used computers
United States	HR 2284: Responsible electronics recycling act, 2011	Banned the export of WEEE items: personal computers, televisions, printers, Xerox, phones, CRTs, batteries containing Pb, Cd, Hg, Cr, Be, and organic solvents
Hong Kong	Advice on the movement of UEEE, 2011	Legislative control on UEEE
Nigeria	Guide for importers of UEEE into Nigeria, 2011	Import of WEEE banned with compulsory registration of importers
India	E-waste rule, 2011	Implemented EPR rule

the environment. Effluents are generated without adequate treatment or safeguards. The health effects associated with e-waste are shown in Table 3.4.

Recently, global initiatives have supported formal recycling and Figure 3.7 exhibits the process flow diagram for e-waste recycling. The first step is to dismantle the e-waste and then shred it in a shredder machine. Then it will pass through a magnetic separator to segregate the magnetic and non-magnetic metals. The plastics present in the waste are collected and sent to the authorized plastic recycler. Around 40% of e-waste contains components like PCBs, which are directly treated for precious metal recovery by pyrometallurgy or hydrometallurgical processing routes. Cathode ray tube (CRT) recycling follows cutting,

Table 3.4 Toxicity of the substances in e-waste and their impact on public health.[13]

Toxic Metal	E-waste Components	Limit, ppm	Disease Caused by Exposure above Permissible Limit
Ag*	Ceramic capacitors, switches, batteries	5.0	Blue pigments on body, damages brain, lung, liver, and kidney
As**	Gallium arsenide is used in LEDs	5.0	Skin disease, lung cancer, impaired nerve signalling
Ba**	Electron tubes, lubricants, fluorescent lamps, CRT guns	<100	Brain swelling, muscle weakness, damage to the heart
Be**	Power supply boxes, motherboards	0.75	Lung cancer, berylliosis, skin disease, carcinogens
Br**	PCBs, casings, PVC cables	0.1	Thyroid gland damage, hormonal issues, skin disorder, DNA damage, hearing loss
Cd**	PCBs, batteries, CRTs, semiconductors, infrared detectors, printer inks, toners	1.0	Pose a risk of irreversible impacts on human health particularly the kidney
CN**	PCBs	<0.5	Cyanide poisoning, >2.5 ppm may cause coma and death
Cr(VI)**	Plastic computer housing, cabling, hard discs, colorants in pigments	5.0	DNA damage, permanent eye impairment
Hg**	Batteries, liquid crystal displays, switches, backlight bulbs or lamps	0.2	Damages brain, kidney and fetuses
Li*	Mobile phones, batteries	<10#	Diarrhea, vomiting, drowsiness, muscular weakness
Ni*	Batteries, semiconductors, CRTs, PCBs	20.0	Allergic reaction, bronchitis, reduces lung function, lung cancers
Pb***	Transistors, LEDs, lead-acid batteries, solders, CRTs, PCBs, florescent tubes	5.0	Damages brain, nervous system, kidney, and reproductive system, causes acute and chronic effects on human health
Sb**	CRT glass, plastic computer housing, solder alloys	<0.5	Cancer, stomach pain, vomiting, diarrhea, stomach ulcer
Se**	Fax machines, photoelectric cells	1.0	High concentration causes selenosis

(Continued)

Table 3.4 (*Continued*)

Toxic Metal	E-waste Components	Limit, ppm	Disease Caused by Exposure above Permissible Limit
Sr***	CRTs, batteries	1.5	Somatic as well as genetic changes due to this cancer in bone, nose, lungs, and skin
Zn**	Batteries, luminous substances	250.0	Nausea, vomiting, pain, cramps and diarrhea
CFCs**	Cooling units, insulation foams	<1.0 for 8 h/day	Impacts on the ozone layer which can lead to greater incidence of skin cancer
PCBs**	Transformers, capacitors, condensers	5.0	Cancer in animals, can lead to liver damage in humans
PVC**	Monitors, keyboards, cabling, plastic computer housing	0.03	Hazardous and toxic air contaminants, release of HCl causes respiratory problems

*Critical; **Hazardous and toxic; ***Radioactive waste; #Limit in serum/blood. CFC: chlorofluorocarbon; PVC: polyvinyl chloride.

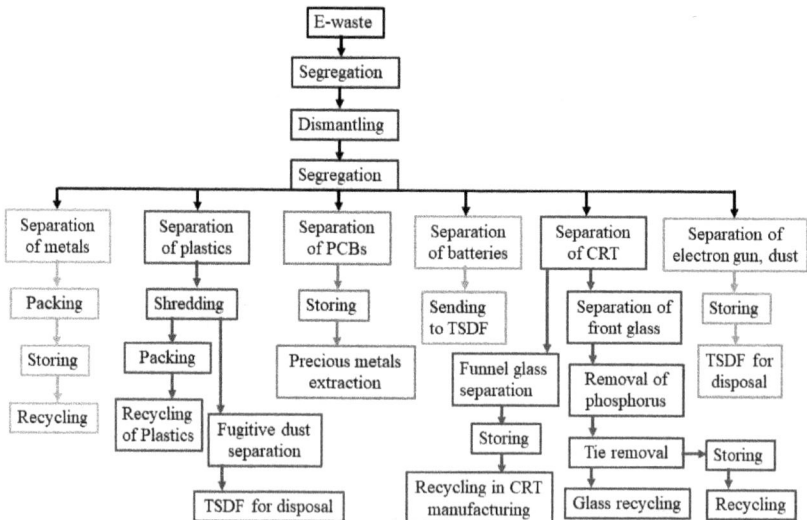

Figure 3.7 Steps of formal recycling of e-waste.[14] TSDF: Treatment storage disposal facility.

shredding, shadow mask removal, crushing, washing, and re-melting of the crushed/fused glass (cullet) process. An acid or alkaline washing is provided on the surface-coated layer of the panel and funnel; however, drum washing uses an eco-friendly apparatus for this operation. The sludge obtained from glass washing is a secondary waste that again gets recycled. CRT recycling can remove hazardous elements like Hg from the process materials; however, some hazardous wastes such as electron guns must be referred to the authorized treatment units for final disposal by following the pollution control board norms.

The benefits of formal recyclers are eminent: recovery of the valuable (Co, Cu, Ni, Li, etc.) and precious metals (Au, Ag, Pt, Pd, Rh, etc.) from the PCBs of computers and mobile phones. The shredded PCBs of size 5 mm are fed to Cu/Pb smelter in a fixed quantity mixed with the fresh ore/concentrate charge for the recovery of copper and precious metals. 95% selective copper recovery over precious metals is achieved using ammonia-based and subsequent sulfuric acid leaching. However, there are some disadvantages of this process such as air pollution and losses of heavy (Zn, Sn, Pb) and noble metals, therefore mechanical pretreatment in combination with the hydrometallurgical recovery of precious metals is recommended. The search for suitable greener reagents is also ongoing to introduce more sustainability to the recycling process, though none has yet to be adopted in commercial applications. Finally, formal e-waste recycling contributes to greener economic growth as presented in Table 3.5.

Table 3.5 Reduction of greenhouse gases in the formal recycling process.[13]

Resource Recovery (× 1000 tons)		CO_2 Reduction on Recycling (× 1000 tons)	
Ferrous metal	404	Refrigerators	256
Non-ferrous metals	37	Washing machines	196
Plastics	180	Televisions	100
Others	93	Air-conditioners	16
Total	720	Total	620

Accordingly, reverse logistics has been implemented to bring sustainability to developed nations, which is now adopted by developing nations. Three major steps of the reverse logistic procedure for e-waste recycling are shown in Figure 3.8. Unfortunately, developing nations are struggling to implement all recommended steps of recycling and disposal because of limited infrastructure accessibility, technological access, and investments. Figure 3.9 advocates the setting up of collection and screening centers by metropolises based on the number of inhabitants. It shows the technical and logistic combination of the best pre-processing of e-waste in developing countries to dismantle e-waste manually and the best end-processing to treat hazardous and complex fractions of e-waste. In addition, cleaner production design, EPR, standards and labeling, product stewardship, and remanufacturing are some of the other good practices implemented by various nations to deal with e-waste and bring sustainability.

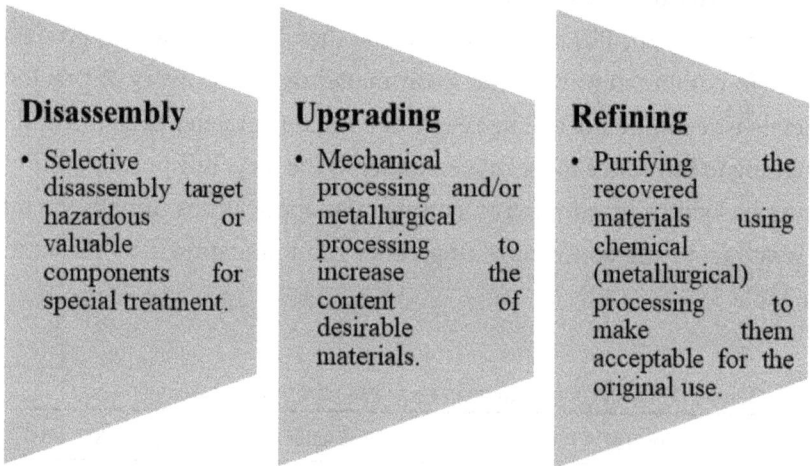

Disassembly

- Selective disassembly target hazardous or valuable components for special treatment.

Upgrading

- Mechanical processing and/or metallurgical processing to increase the content of desirable materials.

Refining

- Purifying the recovered materials using chemical (metallurgical) processing to make them acceptable for the original use.

Figure 3.8 Major steps of the reverse logistic process in e-waste management.

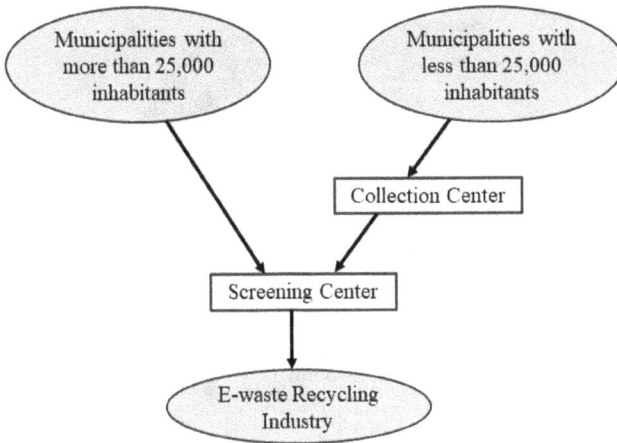

Figure 3.9   Proposed collection center and screening center based on population size.[7]

3.3.3 E-waste Recycling, Circular Economy and Environmental Implications

As defined in Chapter 1, recycling and the circular economy have the potential to reduce the net consumption of metals by promoting resource circulation, which also helps to minimize waste generation. It includes multi-R concepts such as reduce, reuse, refuse, recycle, recovery, rethinking, redesign, etc., that highlight the social, environmental, and economic aspects.[15] The concept of a circular economy in e-waste management is shown in Figure 3.10.

The environmental benefits of metal recycling from e-waste can be clearly demonstrated by examining the quantity of metals recovered and their reintegration into the circular economy, especially when compared to the environmental impact of primary ore mining and water usage. For example, printed circuit boards (PCBs) typically contain around 13% copper. Assuming a 20% recycling rate of current e-waste and a 92% copper recovery efficiency, it has been estimated

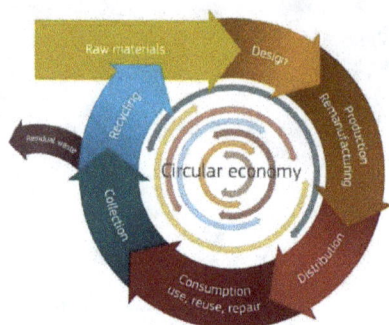

Figure 3.10   Circular economy for e-waste management.

Table 3.6   Environmental impact comparison between microbial recycling of e-waste and primary copper ore mining.

Recycled Quantity	Standard Data of Primary Mining of Copper			Copper Recovery from the Urban Mining of E-Waste		
	CO_2 Emission	Water Use	Water Withdrawal	CO_2 Emission	Water Use	Water Withdrawal
51,833 tons	3.7 kg/kg Cu	69.2 m^3/t	245 m^3/t	173.98 million kg	23.58 million m^3	12.7 million m^3

that 51,833 tons of copper can be recovered.[18] This recovered copper would offset the need to mine approximately 2.35 million metric tons of primary ore, assuming a standard copper concentration of 2.0 kg per ton of ore. Avoiding this scale of mining would result in a significant reduction in CO_2 emissions, calculated using the standard emission factor of 3.7 kg CO_2 per kg of primary copper produced. As shown in Table 3.6, the recovery of 51,833 tons of copper could prevent the release of approximately 173.98 million kg of CO_2, along with saving

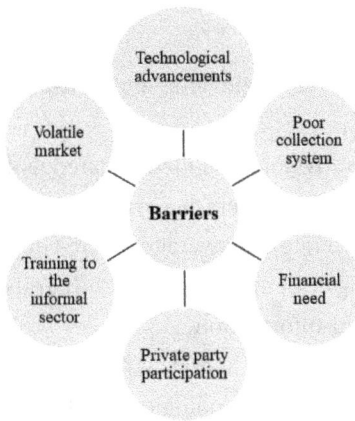

Figure 3.11   Barriers to sustainable e-waste management.

23.58 million m³ of water use and 12.7 million m³ of water withdrawal. These values highlight the considerable environmental advantages of urban mining and recycling practices over traditional mining and extraction processes.

However, various barriers have been identified that show the implementation limit for material recovery through e-waste recycling as shown in Figure 3.11. It is recommended that the informal sector (especially rag pickers) should have advanced training for collection, dismantling, and refurbishing. They should also get incentives and financial support to attract private party participation, stringent regulations for importing e-waste, monitoring and enforcement of existing regulations, international collaborations, and decentralized e-waste refurbishment initiatives. These initiatives can help to build a sustainable e-waste model for developing nations.

References

1. https://www.iea.org/reports/global-ev-outlook-2020.

2. https://www.frame.lneg.pt/.

3. https://www.automotivelogistics.media/materials-handling/running-out-of-juice-will-demand-for-battery-materials-outstrip-supply/38647.article.

4. https://copperalliance.org.uk/coverage-future-copper-demand/.

5. https://www.usgs.gov/centers/national-minerals-information-center/copper-statistics-and-information.

6. Abdelbaky, M., Peeters, J. R. and Dewulf, W. (2021) On the influence of second use, future battery technologies, and battery lifetime on the maximum recycled content of future electric vehicle batteries in Europe, Waste Management, 125, 1–9.

7. Srivastava, R. R. and Pathak, P. (2020) Policy Issues for Efficient Management of E-waste in Developing Countries, in *Handbook of Electronic Waste Management*, pp. 81–99, Elsevier.

8. https://www.pv-magazine.com/2019/11/07/e-waste-company-opens-battery-recycling-plants-in-singapore-and-france.

9. Choi, Y. and Rhee, S. W. (2020) Current status and perspectives on recycling of end-of-life battery of electric vehicle in Korea (Republic of), Waste Management, 106, 261–270.

10. https://eur-lex.europa.eu/legal-content/EN/TXT/PDF/?uri=CELEX:32002L0095&from=EN.

11. https://noharm-uscanada.org/sites/default/files/documents-files/84/Exporting_Harm_Trashing_Asia.pdf.

12. http://ewastemonitor.info/.

13. Pathak, P., Srivastava, R. R. and Ojasvi (2017) Assessment of legislation and practices for the sustainable management of WEEE in India, Renewable & Sustainable Energy Reviews, 78, 220–232.

14. Pathak, P., Srivastava, R. R. and Ojasvi (2019) Environmental Management of E-waste, in *Electronic Waste Management and Treatment Technology*, pp. 103–132, Elsevier.

15. https://www.unido.org/sites/default/files/2017-07/Circular_Economy_UNIDO_0.pdf.

16. Achzet, B., Reller, A., Zepf, V., Rennie, C., Ashfield, M. and Simmons, J. (2011) *Materials Critical to the Energy Industry*, BP Plc.

17. Ilyas, S., Kim, M. S., Lee, J. C. et al. (2017) Bio-reclamation of strategic and energy critical metals from secondary resources, Metals, 7(6), 207.

18. Ilyas, S., Srivastava, R. R., Kim, H., Das, S. and Singh, V. K. (2021) Circular bioeconomy and environmental benignness through microbial recycling of e-waste: A case study on copper and gold restoration, Waste Management, 121, 175–185.

CHAPTER FOUR
SUSTAINABLE FOOD AND LOSS MANAGEMENT

4.1 Introduction

Waste generation in the food supply chain has become a serious global concern. Due to population expansion and increasing scarcity of natural resources, food wastage has resulted in not only increased economic expenses but also social costs and a negative impact on our environment.

To decrease food waste and its negative consequences, it is critical to have a clear definition of "food waste" and a specific and accurate classification of the various types of food waste. This chapter discusses the fundamentals of our food supply system, clearly delineating food loss and waste. Then, using facts and numbers, the sustainability implications of the food supply chain are examined, revealing how our present food production and consumption model affects our environmental, social, and economic sustainability.

The circular economy is a relatively new idea that embraces the concepts of reduce, reuse, and recycle. These principles are discussed and applied to our current model of food production and consumption, demonstrating how to create a long-term sustainable food supply chain. Following that, a complete framework for achieving an efficient and sustainable food supply chain is described, including sustainable boundary conditions, life cycle assessment, and circular principles.

4.2 Food Loss and Waste

Food loss and waste is a global challenge in the 21st century that affects both developed and developing nations. Food is wasted at every stage of the supply chain, from production through transportation, storage, consumption, and post-consumption. Food that is lost or discarded represents a missing chance to feed the global rising population.

Figure 4.1 depicts the food loss and waste caused along the supply chain. Food is lost due to improper handling during the production or harvesting phases (e.g., harvesting using machinery often leads to the loss of grains). It accounts for nearly 24% of food loss in the entire supply chain. Post-harvest and storage losses account for another 24%, and

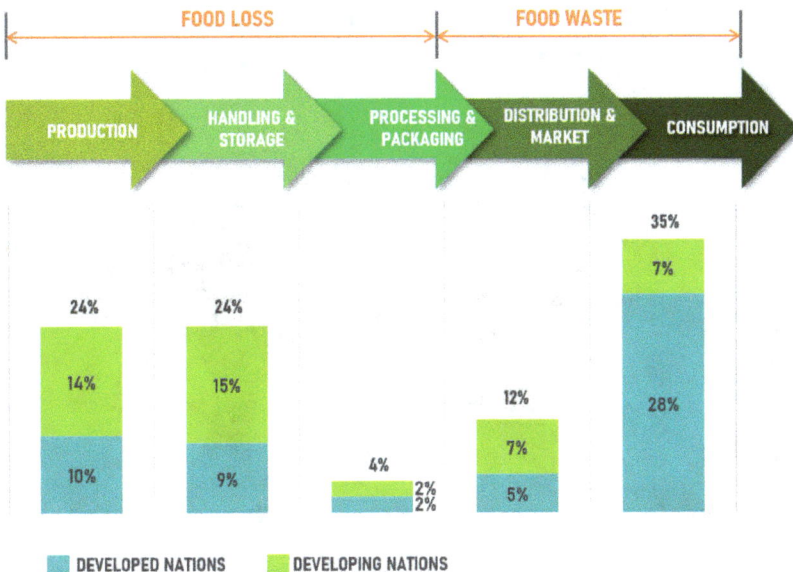

Figure 4.1   Food supply chain diagram illustrating loss and waste.

are caused by product spillage and mistreatment, as well as a shortage of cold storage facilities. Additionally, there are losses, as little as 4%, associated with storage and transportation between the farm and the distribution facility. Also, during the distribution and market phase, we tend to lose 12% of the food. Significant food loss happens during the consumption phase, contributing up to 35%. This phase is responsible for massive food waste generation in the entire supply chain.[1,2]

Figure 4.2 depicts the alarming statistics about population growth concerning the current food supply and management. In 2020, the urban population accounts for nearly 4.3 billion people, occupying approximately 56.2% of the global landscape. Over 2.4 billion tons of food were consumed by them, with 0.55 billion tons of food being wasted, or approximately a quarter, which can feed 1 billion people. If present trends continue, by 2050, 68% of the world's population would live in cities,

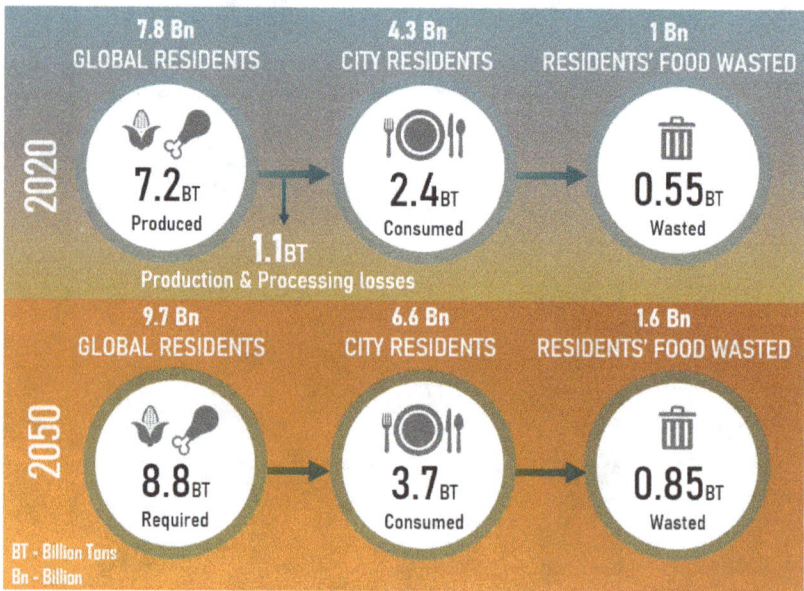

Figure 4.2   Linear economy model of food production and consumption in line with a growing population.

capable of producing 0.85 billion tons of food waste.[3] This scenario illustrates the consequences of a continued belief in the linear food production and consumption paradigm, based on the premise that resources are limitless and renewable in a short period of time.

As illustrated in Figure 4.1, our food supply chain is divided into five different phases, and we know the proportion of food loss/waste created in each phase. However, we need to understand the underlying factors in order to come up with targeted solutions (Table 4.1). For instance, during the production phase, farmers often overproduce crops, resulting in low market demand, economic hardship, and depreciation of food products. Developing nations also often lack the infrastructure to store the food products efficiently, resulting in significant losses along with other factors such as loss during transportation. Contributing reasons for other phases portray our current food production and consumption as very inefficient and require sustainable intervention to produce

Table 4.1 Underlying factors contributing to food loss and waste along the supply chain. Reproduced with permission.[2]

Production	Handling & Storage	Processing & Packaging	Distribution & Market	Consumption
› Infrastructural limitation	› Degradation and spillage of product composition	› Unavoidable losses	› Transport conditions	› Composition unit and size of household
› Over production	› Loss during transportation from farm to distribution	› Technical malfunctions	› Contamination of transport	› Income group
› Harvesting timing and method	› Storage infrastructure	› Methods and changes in processing lines	› Transport and market facilities	› Demographic and culture
› Pesticides and fertilizers		› Contamination in processing	› Road and distribution vehicles	› Individual attitude
› Economic problems		› Legislation restrictions	› Packaging management	› Cooking practices and methods
› Quality standards and norms		› Packaging systems	› Commercial conditions	

significant savings to keep up with the growing population. After addressing the environmental impacts of our existing food production and consumption model, we will discuss sustainability interventions.

4.3 Sustainability Implications of Current Food Production

4.3.1 Economic Implications

Food loss and waste have many negative economic impacts on monetary costs across the globe. For example, the United Nations Food and Agriculture Organization (FAO) estimated that the economic costs of food wastage amount to about USD 1 trillion each year.[4] If we split the data by nation, we arrive at the following statistics. 30% of all food consumed in the United States is discarded during the consumption stage, resulting in a food loss of US$48.3 billion per year. Similarly, food waste occurs during transit and in supermarkets, amounting to around US$90–100 billion per year. China discards over US$32 billion worth of food each year. Each year, India loses and wastes 580 billion rupees worth of food. $10.5 billion was spent in Australia on goods that were never used or discarded.[5] This equates to almost $5,000 per capita every year. Annually, the EU generates around 88 million tons of food waste, with related expenses estimated at 143 billion euros.[6]

4.3.2 Environmental Implications

Food production contributes 26% of global greenhouse gas (GHG) emissions.[7] However, nearly 6% of GHG comes from food loss and waste, which is equivalent to 2.1 Bn tons of CO_2, according to the Carbon Brief organization in 2021. As seen in Figure 4.3, food loss and waste in the supply chain involve 173 billion cubic meters of water each year, accounting for 24% of all water used in agriculture. Each year,

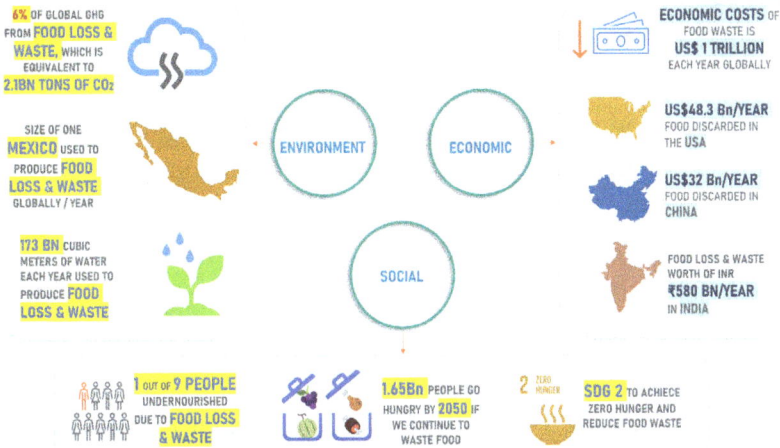

Figure 4.3 Sustainability (environmental, economic and social) implication of our food production and consumption model.

198 million hectares/year of farmland are employed to cultivate this lost and wasted food, an area about the size of Mexico, with an associated 28 million tons of fertilizer.[1]

4.3.3 Social Implications

According to a United Nations survey, the global population in 2020 was 7.8 billion, and it is estimated to grow by about one billion by 2025 and exceed 9.8 billion by 2050.[8] The FAO estimates that around a third of the food meant for human consumption is lost or wasted across the globe, resulting in one out of every nine people being undernourished.[4] As seen in Figure 4.2, the urban population in industrialized countries has a greater tendency to waste food, accounting for about 0.55 billion tons of food loss during the consumption phase, which is enough to feed 1 billion people. If the current rate of food loss and waste continues, an estimated 1.6 billion people would go hungry or be undernourished by 2050.[3] To that end, the current draft of the United Nations Sustainable Development Goals includes a targeted goal #2 titled 'Zero Hunger,' which focuses on halving food waste per capita during the consuming phase by 2030.[9]

Figure 4.3 illustrates the detailed sustainability implications of our existing food loss and waste, depicting that our whole food supply chain is truly unsustainable. Along with these sustainability implications, we are also facing other consequences of improper food waste management, including water pollution, nutrient leaching, and use as a breeding ground for illnesses and other health threats to humans and animals.

According to 2012 World Bank estimates, over 50% of food waste is disposed of in open landfills, while developing countries are still unaware of proper food waste disposal, with between 13% and 33% of waste being discarded openly. These landfills and dumpsites pose numerous health risks to the people who live and work nearby. For instance, organic waste at dumpsites provides a breeding ground for mosquitoes and flies, increasing the danger of food-borne illness transmission.[2]

4.4 Embracing Circular Principles in Food Waste and Loss

So far, we have seen how food is wasted and lost along the whole supply chain, which has implications for long-term sustainability. As seen in Figure 4.4, researchers have highlighted many areas for improvement to help decrease the global food waste and loss crisis. In this sense, the circular economy is a developing solution founded on the concepts of *reduce*, *reuse, and recycle* to address global concerns such as food waste and loss.

4.4.1 *Reduce*

The value of food cannot be overstated, and it is a resource that should not be mismanaged. The food waste hierarchy prioritizes preventing food waste generation (Figure 4.4). To begin, we must identify precisely

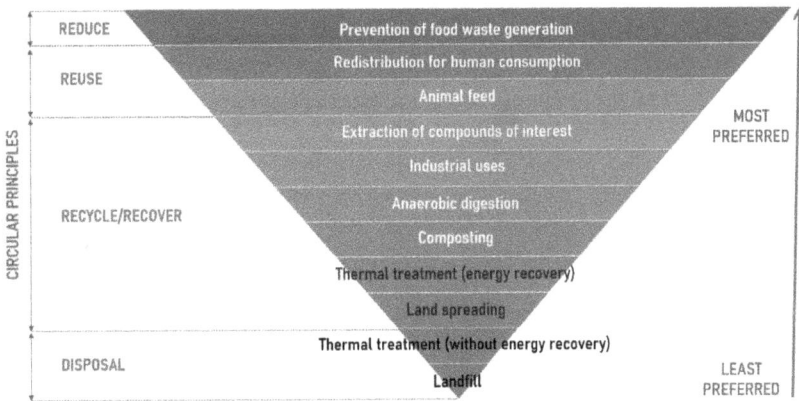

Figure 4.4 A waste hierarchy for excess food and food waste is being developed in accordance with circular concepts. Adapted with permission.[10]

where food is wasted or lost within the hierarchy, in order to apply circular concepts. As discussed in this chapter (Figure 4.2), food is mostly wasted in two stages: production and distribution in developing countries and consumption in developed countries.

We cannot reduce food consumption, due to population growth. However, the principles of circular economy suggest that we can design a system to measure, and establish targeted objectives for reducing, food loss and waste. Following that, increased investment in knowledge and infrastructure for post-harvest losses reduction in underdeveloped and developing nations should be encouraged, along with the establishment of entities in food waste reduction in developed countries.[1]

4.4.2 Reuse

The second strategy of the circular economy idea is reuse, which refers to the process of reusing food waste generated in the system. This method is most effective at the consumption stage (Figure 4.5), especially when food with a short shelf life is supplied for human or animal consumption. For instance, bread is the most squandered product in

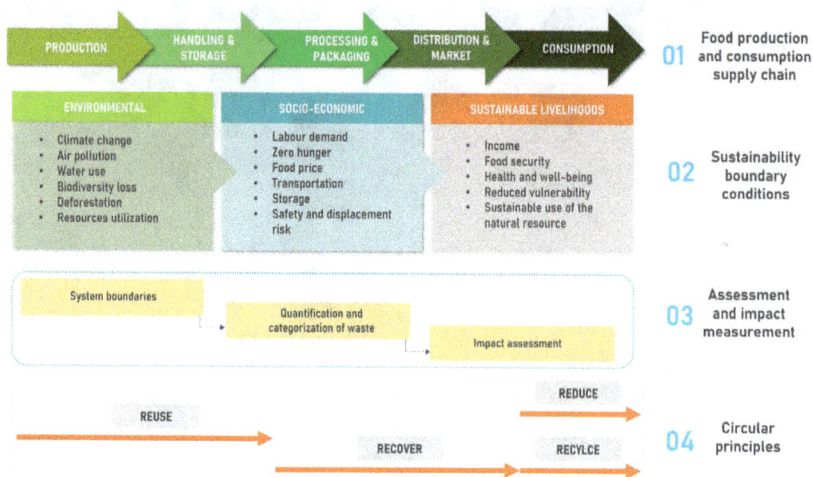

Figure 4.5 Sustainable food supply chain framework. Adapted with permission.[4,10]

the United Kingdom, accounting for 900,000 tons per year,[11] or about 42%.[12] A recent study indicates that this sort of food waste, particularly dated bread, may be repurposed as a feedstock for brewing industries or utilized for animal feed through food waste valorization.[13]

4.4.3 Recycle/Recover

The last phase of circular principles is recycling and recovering useful components of food waste. For instance, fats, proteins, and bioethanol may be recovered and utilized for various applications. Over the past several years, there has been an increase in scientific interest in recovering valuable substances from food waste. These extraction techniques require sophisticated equipment, however, and industrial-scale technologies are not immediately available due to their high cost of operation. Nevertheless, this sector is growing significantly.[10]

Anaerobic digestion and composting are two processes that assist in recycling post-consumer food waste, mainly vegetables or meat. Anaerobic digestion is the process of decomposing organic matter with the help of microbes without oxygen. This procedure requires an alkaline pH, a controlled temperature, and minerals like nitrogen, phosphorus, and potassium. This method has been shown to significantly minimize the health hazards associated with improper waste management and to assist in the recovery of energy from food waste. Anaerobic digestion recovers 60% more energy than direct combustion and contributes to the production of a nutrient-dense fertilizer as a byproduct.[10]

Biological composting is a simple process that uses microorganisms to turn food waste into nutrient-rich organic materials. Composting may be done on a small scale, such as home composting, or a large scale, such as in-vessel composting, for huge amounts of food waste undertaken by governments. This method can produce high organic matter compost, which may be used to increase plant yields and soil fertility.

4.5 Sustainable Food Supply Chain

To create a more sustainable food supply chain, it is critical to reduce food waste at all levels. Though several organizations and research groups have achieved significant strides in recent years, reducing environmental and social consequences as a result of food waste and loss, much remains to be done.

As previously noted, our food system is inefficient in terms of food waste and loss management. In addition to the economic costs, there have been consequences on the ecological and social aspects, reflecting the unsustainability of our current model. As a result, a comprehensive framework is necessary to accomplish the finest sustainable practices, as illustrated in Figure 4.5. When evaluating our food supply chain, for example, a set of sustainable boundary criteria must be established, together with environmental elements and socio-economic features that contribute to sustainable lifestyles.

As seen in Figure 4.5, after the sustainable boundary conditions are defined (02), the following step (03) is to assess and quantify the impacts on sustainability using life cycle assessment. When establishing the system's boundaries, it is critical to specify which components of the supply chain will be studied. For instance, if we study 'food loss', we must consider our boundary conditions, which include food production, harvesting, transportation, storage, packing, and distribution. Similarly, while analyzing 'food waste', we must examine the consumption phase, which includes food waste generated in homes, restaurants, incineration, landfills, and sewage disposal.

Quantification and classification include identifying the types of food products that go unused or are discarded. Generally, fruits and vegetables create greater waste; for example, 44% of waste occurs during processing and packing.[10] Thus, in establishing the life cycle assessment, it is necessary to consider the repurposing of waste connected with these edible goods into animal feed. Similarly, although meat products tend to generate less waste across the supply chain, their environmental effect is enormous. For example, 60 kg and 24 kg of CO_2 are emitted into the atmosphere to create one kilogram of beef and mutton, respectively.[14] Hence, detailed impact assessment methods associated with each food waste must be formulated to identify the sustainability impact.

Preventing food waste is the ideal alternative, which may be accomplished by adhering to circular principles. As indicated in Section 4.4, food-related losses, such as excess food before packaging, in the supply chain may be reused for animal needs. Following that, surplus food in the supply chain may be collected for recycling or repurposing during the processing and distribution stages via appropriate life cycle assessments. Recycling or upscaling food waste is another sustainability option during the consumption stage. While food waste during consumption is inevitable, technologies such as anaerobic digestion and composting may be used to treat all biodegradable waste and keep them out of landfills.

4.6 Conclusion

Food supply chains in both developing and developed countries suffer enormous losses and waste. It is vital to minimize the environmental, social, and economic costs associated with food loss and waste. As a result, new methods and sustainable management techniques are needed to curb the rise of food-related unsustainable practices. Unfortunately, it is very difficult to quantify food loss and waste across the globe due to our different practices and management strategies. We addressed in this chapter how our existing food production and consumption model has affected our environmental, social, and economic sustainability, preventing us from achieving a sustainable food supply chain. To assist in addressing food loss and waste concerns, a common framework is required that can be used to investigate different forms of waste at various points throughout the supply chain, assess their sustainability impact, and provide guidance on the approach or technology to apply to manage these waste issues. Using the framework proposed in this chapter, which incorporates sustainable practices while measuring food loss and waste via the life cycle assessment method and circular principles, it would be easier to quantify, monitor, and prevent food waste, thereby ensuring a long-term sustainable food supply chain. In the long run, waste treatment technologies, novel research methodologies such as converting food waste to activated carbon to absorb carbon emissions, and life cycle assessments will provide a clear path forward for resolving the food waste problem.

References

1. Schuster, M. and Torero, M. (2016) Toward a Sustainable Food System: Reducing Food Loss and Waste, in *2016 Global Food Policy Report*, pp. 22–31, International Food Policy Research Institute (IFPRI).
2. Singh, K. (2020) Sustainable Food Waste Management: A Review, in *Sustainable Food Waste Management*, pp. 3–19, Springer.

3. Selvan, R. T. and Ramakrishna, S. (2022) *Sustainability for Beginners: Introduction and Business Prospects*, World Scientific.

4. https://www.fao.org/3/i3991e/i3991e.pdf.

5. https://www.stopfoodlosswaste.org/.

6. https://ec.europa.eu/food/safety/food-waste_en.

7. https://ourworldindata.org/food-waste-emissions.

8. https://www.un.org/en/desa/world-population-projected-reach-98-billion-2050-and-112-billion-2100.

9. https://sdgs.un.org/goals/goal2.

10. Garcia-Garcia, G., Woolley, E. and Rahimifard, S. (2015) A framework for a more efficient approach to food waste management', International Journal of Food Engineering, 1(1), 65–72.

11. https://www.ecoandbeyond.co/articles/much-bread-waste-uk/.

12. https://bakeryinfo.co.uk/finished-goods/bread-tops-food-waste-list-finds-study/619896.article.

13. Verni, M., et al. (2020) Wasted bread as substrate for the cultivation of starters for the food industry, Frontiers in Microbiology, 11, 293.

14. https://ourworldindata.org/food-choice-vs-eating-local.

CHAPTER FIVE
SUSTAINABILITY OF PLASTIC MATERIALS

5.1 What is Plastic?

Plastic is a synthetic/semi-synthetic, highly versatile material that has a long hydrocarbon chain of polymer that also includes oxygen, nitrogen, or sulfur. It is derived from the

> Plastic is defined as a material that contains an essential organic substance of large molecular weight.
>
> All plastics are polymers, but not all polymers are plastics.

Greek word *plastikos*, which means fit for molding or plasticity. Polymers are natural or synthetic materials made up of chains of monomers.

Petro-based plastics are prepared from the refining process of non-renewable organic fossil fuels such as crude oil, coal, minerals, and natural gas. It accounts for 4–6% of global consumption of fossil fuels. Naphtha obtained from petroleum distillate is an important raw material for plastics manufacturing. Based on boiling point, naphtha is classified into light and heavy.

Two processes, i.e., polymerization and polycondensation, are used to make plastic products in the presence of a suitable catalyst. Initially, a low-molecular-weight compound (monomer) is formed, which then binds to other monomers to yield a high-molecular-weight compound (polymer). Plastics are classified based on thermal behavior and size (Figure 5.1). Thermosets, thermoplastics, and elastomers are classified based on thermal behavior, wherein thermosets are very hard and have a branched molecular chain structure. Once thermosets are molded, they cannot be further softened or remolded even if heat is applied. Thermoplastics have linear or branched molecular structures and are softened at ordinary temperatures. The chemical characteristics of thermoplastics do not change with heating or cooling. Finally, elastomers are cross-linked polymer structures with looser mesh than thermosets. It is elastic but once molded it will not change shape again upon heating. An example is automobile tires.

Based on size, we can alternatively classify plastics into micro, meso, and macroplastics. However, based on consumption, mostly seven types of plastics are used in different sectors, as shown in Figure 5.1.

Plastics are robust and highly versatile materials. It is a poor conductor of heat, corrosion-free and easily shaped and sized. Owing to these properties, plastics are perfect for an extensive range of customer and industrial applications. Therefore, many different sectors (depicted in

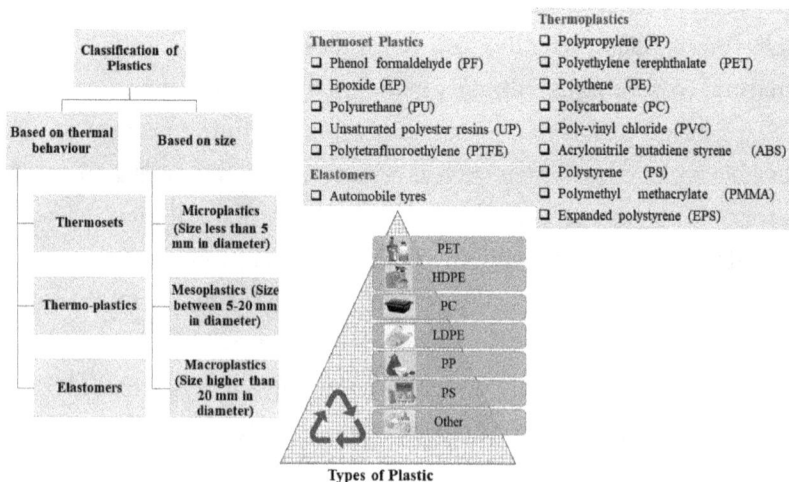

Figure 5.1 Classification and examples of petro-based plastics.

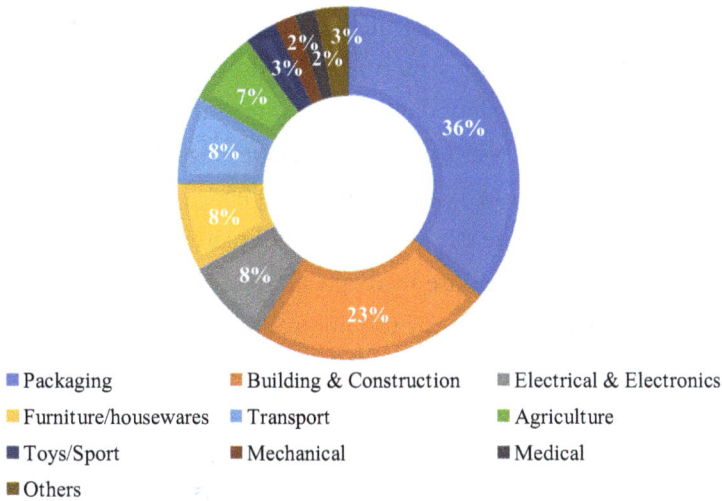

Figure 5.2   Sectors that use plastic products.

Figure 5.2) use products made of plastics, and their wide applications have made our lives easier, more comfortable, cleaner, and enjoyable.

Population growth, urbanization, and industrialization have increased the consumption of plastics and enhanced businesses and economies. However, its ready availability and low price have encouraged disposal after its end of life. Plastic consumption has steadily increased for a few decades, yet there has been no proper management after its end of life. It is disposed of either directly in landfills as waste or dumped into the ocean. The Organization for Economic Co-operation and Development stated that globally, we had produced ~359 million tons of plastics in 2019. Of this, 49% was produced in the Asia-Pacific region itself, and 38% consumed therein.[1] However, ~8 million tons of plastics are dumped in the ocean every year, with most flowing in from river conduits.

5.2 Mismanagement of Plastics and its Impacts

The plastic waste generated worldwide follows the linear economy model, i.e., take-make and dispose of, where disposal is done directly to landfills and oceans or openly incinerated. However, due to the complex chemical composition, structure, and addition of antioxidants and stabilizers, plastic has a slow degradation rate and thus long life expectancy. For instance, it was found that plastic bags and plastic bottles take 500–1,000 years and 70–150 years, respectively, to degrade in the natural environment. The degradation rate of plastic waste can be accelerated by pre-treatment before its disposal; the difference with the non-treated will vary depending on the environmental conditions such as landfill/soil, marine, biological degradation, and sunlight as shown in Figure 5.3.

Globally, ~380 million tons of plastic waste were produced in 2015 and is expected to reach 1,200 million tons by 2050.[2] While ~16% were recycled, 24% were incinerated, and the rest hoarded in landfills or thrown away in an open natural environment.[3] Plastics contain several organic pollutants such as PoPs (persistent organic pollutants), polychlorinated biphenyls, polycyclic aromatic hydrocarbons, pesticides and heavy metals and cause several health and environmental impacts when thrown improperly as shown in Figure 5.4.

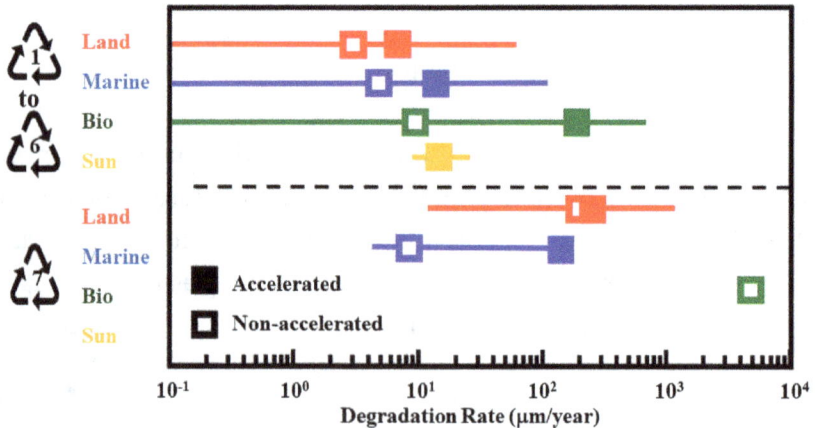

Figure 5.3 Degradation rate of different types of plastics. Adapted with permission.[2]

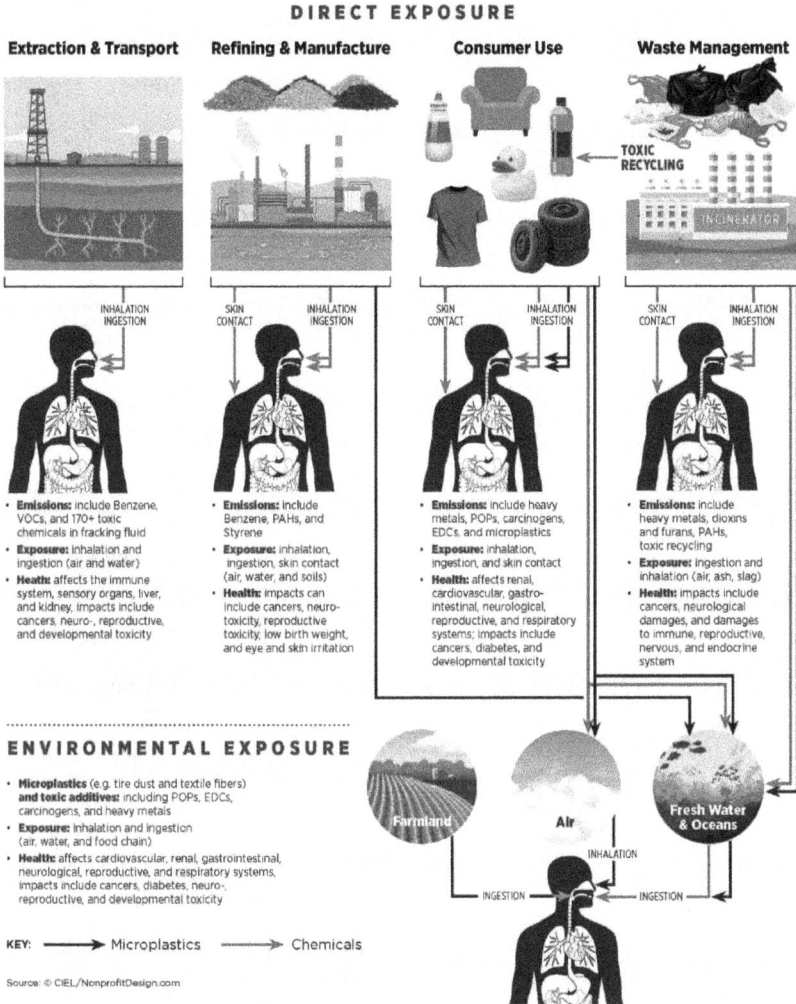

Figure 5.4 Health and environmental hazards associated with plastic pollution. Adapted with permission.[4]

The collection rate for plastic waste has reached up to 100% in developed countries such as Japan, South Korea, and Singapore, and waste segregation at source is common practice in developed nations. However, it is significantly less to moderate in developing countries (~40–80%), and segregation is done mainly by the informal sector. Despite this, a large amount of plastic waste is exported from high-income countries to low-income Asian and Pacific countries. Therefore, the amount of plastic waste is more than its recycling rate.

Numerous plastic materials are recycled across Asia-Pacific. Developed economies, such as Japan, have attained high plastic recycling rates (~20%) in the formal sector. Though the countries in Asia-Pacific claim more than 50% plastic recycling rate, most are carried out in the informal sector and focus on single-use plastic recycling (majority polyethylene terephthalate (PET), polyethylene (PE), and polypropylene (PP)). The low- and middle-income countries such as India, China, and Bangladesh are generally adopting informal, unscientific recycling practices, incineration, or ocean dumping to manage plastic waste. During incineration, greenhouse and organic gas fumes are released into the environment which cause toxic effects on human beings. Moreover, incineration does not completely degrade the plastic. Instead, it breaks down into smaller sizes such as macro, meso, micro, and nano plastics that mix with soil and water to degrade the environment.

The mismanagement of plastic waste causes a potential threat to the terrestrial and marine environment. The United Nations Environment Programme has estimated US$13 billion per year as the economic cost of these environmental impacts. Further, marine plastic debris has also been projected for annual losses of US$622 million for the tourism sector in the Asia-Pacific Economic Area. Further, the Asia-Pacific Economic Cooperation Forum estimated US$1.3 billion in the cost of marine plastics to the tourism, fishing, and shipping industries in the region alone.

5.3 Rules and Regulations on Plastic Management

Due to negative consequences associated with plastic pollution, worldwide rules and policies have been framed, but only developed nations have enforced them successfully. The Montreal Protocol (1989) and

Stockholm Convention (2004) have reclassified hazardous substances present in plastics as chlorofluorocarbons and PoPs that cause adversity in the environment as well as human beings. Furthermore, extended producer responsibility (EPR) was introduced for plastic management, where producers should take responsibility for their products. Moreover, plastic producers should state all constituents in their products and caution for consumers the potential health effects. The multi Rs of the circular economy such as Reduce, Reuse, Redesign, Remanufacture, Recycle, and Recovery must be employed to prevent diversion to land-fills and indiscriminate disposal in the environment. The shift towards a circular economy can create economic opportunities and abate environmental pollution. In this context, plastic waste collection and management are a vital part of the United Nations Sustainable Development Goals (UNSDGs). Good waste management systems not only create healthy and sustainable cities and societies (UNSDG 11) but also safeguard sustainable and effective use of natural resources (UNSDG 12) and reduce marine pollution and preserve life under seas (UNSDG 14).

5.4 Sustainable Recycling of Plastics

Plastic recycling is vital in its management. If they are not recycled at the proper time, then they get mixed with other chemicals or materials and hence become more challenging to recycle. A large number of commercial and innovative recycling technologies are available for plastic recycling, including mechanical, chemical, and thermochemical options. Biological treatment is not successful as they are non-biodegradable, so they do not get decomposed by microbial action. Therefore, it is essential either to recycle plastic waste or use biopolymers or biodegradable polymers.

The ASTM D7209 (2006) has classified four plastic recycling methods as shown in Figure 5.5. It is noticed that there are large variations such as energy use, product outcomes, and emissions due to input parameters.

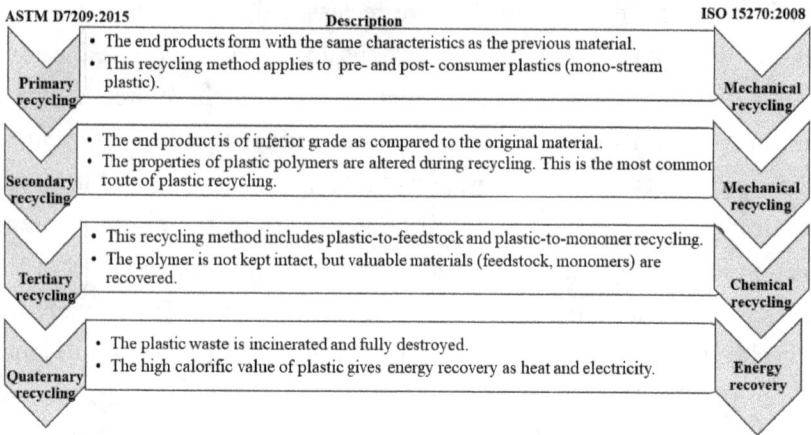

ASTM D7209:2015	Description	ISO 15270:2008
Primary recycling	• The end products form with the same characteristics as the previous material. • This recycling method applies to pre- and post- consumer plastics (mono-stream plastic).	Mechanical recycling
Secondary recycling	• The end product is of inferior grade as compared to the original material. • The properties of plastic polymers are altered during recycling. This is the most common route of plastic recycling.	Mechanical recycling
Tertiary recycling	• This recycling method includes plastic-to-feedstock and plastic-to-monomer recycling. • The polymer is not kept intact, but valuable materials (feedstock, monomers) are recovered.	Chemical recycling
Quaternary recycling	• The plastic waste is incinerated and fully destroyed. • The high calorific value of plastic gives energy recovery as heat and electricity.	Energy recovery

Figure 5.5   Four methods of plastic waste recycling. Reproduced with permission.[5]

The level of progress of the recycling technologies can be defined based on the Technology Readiness Level (TRL) scale: (I) low (1–4), (II) medium (5–7), and (III) high (8–9).

5.4.1 *Primary Recycling (or Closed-loop Recycling)*

The plastic waste is recycled to form an end product without changing the material's characteristics. This type of recycling is considered mechanical recycling of pure plastics. For example, bottle-to-bottle primary recycling includes PP, PET, polystyrene (PS), and polyvinyl chloride (PVC).

5.4.2 *Secondary Recycling (or Open-loop Recycling)*

The plastic waste is recycled but the end product is of inferior grade as compared to the original material. During the recycling process, the plastic polymer altered its properties and therefore it is called open-loop mechanical recycling. In this method, only thermoplastic polymers are used. TRL is very high, and most consumers are using this method.

5.4.3 Tertiary Recycling

This is also called chemical recycling, where waste plastic is chemically converted into monomers. Gasification and pyrolysis are two technologies used for waste plastic that have high TRL. In gasification, the polymer is used as a refuse-derived fuel (RDF) and is converted in a gasifier to syngas with an H_2/CO molar ratio of 2:1. The amount of syngas and the supplementary CO_2 emissions depend on the feed polymer type. On the other hand, in pyrolysis technology, plastics are converted to pyrolysis oil (presumed to correspond to diesel). The energy content of the end product again depends on the calorific value of the feed polymer.

5.4.4 Quaternary Recycling or Incineration

The material is incinerated and destroyed; however, the high calorific value of plastic results in energy recovery as heat and electricity. The TRL of controlled incineration technology is high. The amount of heat and power produced is subject to the polymer's calorific value (energy content) and the configuration of the waste-to-energy plant.

Different recycling technologies are available for plastics through the TRL. These technologies are mainly assessed on different parameters such as operating temperatures, capacity installed, energy inputs, product outcomes, emissions, and scale of operation (Table 5.1). Different plastics melt at different temperatures in underlying environmental conditions. It has been observed that the higher the temperature (>1160°C), the higher will be the polymer breakdown, thereby generating a higher rate of material recovery as in the case of chemical (plasma gasification,

Table 5.1　Different recycling technologies for plastic waste.

Type of Recycling	Advantage	Disadvantage	Feedstock	Capacity, Scale, Location	Temperature (°C)	TRL*
Primary/ Physical	About 80–85% of single uncontaminated plastics get recycled	Multilayer and mixed contaminated plastics remain unidentified due to the difference in shape, size, and color	PET, PP, and high-density PE	1.5 kt/day Commercial in Italy	150–320	6 (medium)
Secondary/ Mechanical	The upcycling process reduces carbon and water footprint, thereby making the process more eco-friendly	In downcycling, there is a degradation of the quality of recycled material	PET bottles	6 t/day Commercial in Singapore	475–520	7 (medium)
Tertiary/Chemical						
Catalytic cracking	Higher oil yield and low tar content	Organic material can block catalyst pores	PP, PE, PS with zeolite catalysts	3 kt/day Commercial in Japan	450–550	9 (high)
Plasma pyrolysis	Product gas with a low tar content	Requirement of high electricity	Mixed plastic waste	14.8 kg/day Laboratory in the USA	4,500– 10,000	4 (low)
Plasma gasification	High purity of the product gas	High operating and maintenance costs	PE, PS	1 kt/day Commercial in Germany	5,000– 15,000	8 (high)
Hydro cracking	High-quality product	Toxic effects from chlorine residues	PVC, PP	240 kg/day Pilot in China	375–500	7 (medium)
Quaternary	It is the most suitable and efficient method for treating inert (non-recyclable) plastics	Emission of toxic gases	PP, PS, PE, multi-layered plastic	10.2 kt/day Commercial in India	>1,160	9 (high)

*TRL-Low: Technologies that have been developed only at the academic or research level. TRL-Medium: It refers to the small (where the capital share is 1–2 crores per annum) and medium-sized (where the capital share is 2–5 crores per annum) companies that are ready with prototypes that can be scaled up easily. TRL-High: It refers to large-sized companies (where the capital share is >10 crores per annum) with proven techno-economic feasibility of the resources.

catalytic cracking) and quaternary recycling, thus making the TRL 8–9 for well-installed large-scale industries. These are the plants where usually more than 3 kilotonnes of plastics are treated per day. In a similar line, primary and secondary recycling operate at comparatively lower temperatures (150–550°C), which increases the contamination from the feedstock and emissions, thereby degrading the quality of the final product. This results in the attainment of a medium level of TRL that is usually in the range of 5–7, for well-established small or medium-sized plants/industries. On the other hand, if the technologies are still in the developmental stage line in laboratory conditions or research and development centers, as in the case of plasma pyrolysis, the TRL is reduced to 1–4.

5.5 Alternatives to Plastics

5.5.1 *Bioplastics*

The ceaseless increase in plastic waste and carbon footprint on plastic products have fueled interest in using alternative environmentally friendly materials. One promising solution is the use of bio-based plastics and composites. It has several advantages over petro-based plastics as shown in Table 5.2.

Bio-based plastics and composites could replace non-degradable plastic polymers. Bio-based materials, *viz.* agricultural and forestry waste, can be used as reinforcement/fillers to develop biocomposites. These materials are renewable and biodegradable and can be an appropriate

Table 5.2   Comparative assessment of petro-based plastics and bioplastics.

	Petro-based Plastic	Bioplastic
Toxic compounds	Included during synthesis	Does not include
CO_2 emission	400 million tons/year	Almost zero emission
Composting process	Takes more than 400 years	Takes a maximum of 180 days
Plastic end product	Threat to organisms	Food to microorganisms

Table 5.3   Classification of bio-based plastics.

Degradability	Source	Types
Biodegradable	It is bio-based and biodegradable.	Bio-PE, bio-PET, bio-polyamide, and bio-poly trimethylene terephthalate
Partially biodegradable	It is bio-based or partially bio-based and non-biodegradable.	Polylactic acid, polyhydroxyalkanoates, polybutylene succinate, and starch blends
Non-degradable	It is fossil-based and non-biodegradable.	Conventional PE, PP, and PET

alternative to petro-based plastics. The classification of bio-based plastics is shown in Table 5.3.

Additionally, the production of bio-based plastics and composites ensures the proper utilization of biomass wastes. The raw materials of biocomposites are classified into two categories: 1st-generation and 2nd-generation materials.

(A) 1st generation
The raw materials are wood, and their use in larger quantities can lead to deforestation and affect biodiversity.

(B) 2nd generation
The raw materials are lignocellulosic wastes collected from food, forest, and agricultural residues. So far, 2nd-generation waste has primarily been used to develop biocomposites, though still it is not commercialized much. Biocomposites can be formed from available plastics (PE from bio-ethanol or bio-ethylene) or can be established from natural renewable sources (polylactic acid from corn starch).

In addition, the biodegradability of oil-based biocomposites can be increased by using bio-fillers. Some polymeric matrices and fillers used for the fabrication of biocomposites are listed in Figure 5.6.

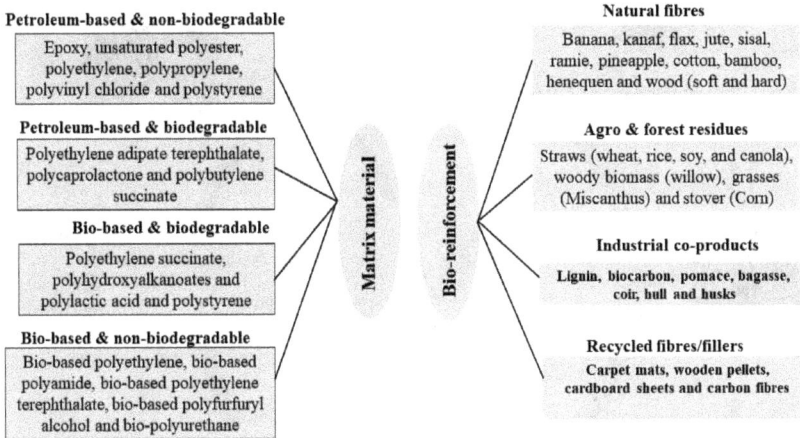

Petroleum-based & non-biodegradable
Epoxy, unsaturated polyester, polyethylene, polypropylene, polyvinyl chloride and polystyrene

Petroleum-based & biodegradable
Polyethylene adipate terephthalate, polycaprolactone and polybutylene succinate

Bio-based & biodegradable
Polyethylene succinate, polyhydroxyalkanoates and polylactic acid and polystyrene

Bio-based & non-biodegradable
Bio-based polyethylene, bio-based polyamide, bio-based polyethylene terephthalate, bio-based polyfurfuryl alcohol and bio-polyurethane

Matrix material

Bio-reinforcement

Natural fibres
Banana, kanaf, flax, jute, sisal, ramie, pineapple, cotton, bamboo, henequen and wood (soft and hard)

Agro & forest residues
Straws (wheat, rice, soy, and canola), woody biomass (willow), grasses (Miscanthus) and stover (Corn)

Industrial co-products
Lignin, biocarbon, pomace, bagasse, coir, hull and husks

Recycled fibres/fillers
Carpet mats, wooden pellets, cardboard sheets and carbon fibres

Figure 5.6   Some matrix and fillers used for the fabrication of biocomposites. Adapted with permission.[6]

The demand for bio-based plastics and composites is estimated to increase at a 11.2% annual growth rate for the period 2017–2030.[6] This demand can only be fulfilled by judicious uses of raw materials and the recycling of bio-based plastics and composites. The recycling process in biocomposites is depicted in Figure 5.7.

The cumulative demand for bio-based plastics has increased its manufacturing. Circular economy tools are thus necessary to sustain the amplified manufacturing volume. They can expand the recyclability, reusability and alter bioplastic waste into valuable products, energy, or secondary materials, which could lead to a reduction in bioplastic waste landfilling. However, the adoption of a circular economy in bioplastic synthesis is challenging and costly due to the intrinsic bonding of the matrix and filler/reinforcement, which makes their separation difficult.

5.5.2 Upcycling of Plastic Waste

Conventional recycling of plastic waste such as mechanical recycling or incineration is called downcycling, where the quality of the end

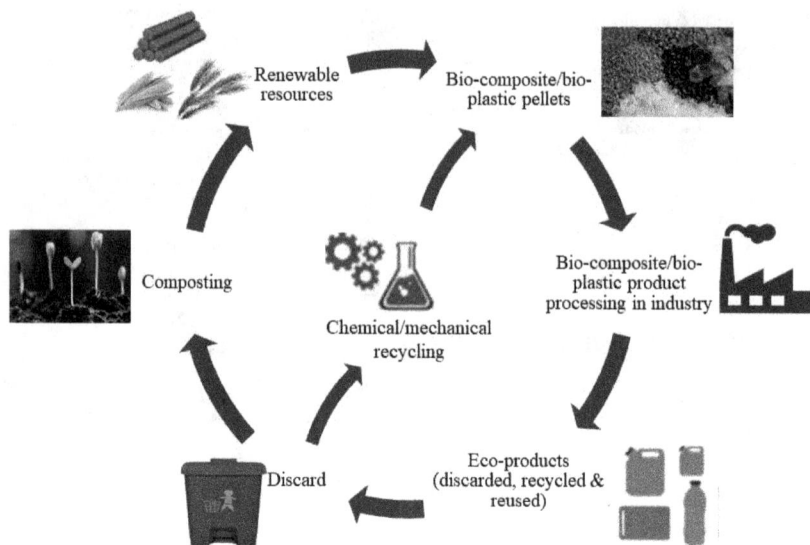

Figure 5.7   Recycling process of bio-based composites. Adapted with permission.[6]

product is downgraded due to loss of its structural integrity. In this context, upcycling or upgrading recycling is one of the emerging alternatives, where waste plastics or low-graded plastics are converted or transformed into higher-value and eco-friendly materials. In this process, for example, plastic bottles can be used/upcycled as flower pots, garden sprinklers, bird feeders, etc. Waste plastic can also be blended with cotton materials or inorganic materials to make clothes and eco-bricks, respectively. Levi has successfully mixed plastic waste into cotton and produced jeans that come from 100% waste materials/garbage. As the textile industry has historically caused huge pollution, upcycling plastic polymers with cotton can reduce environmental impacts and minimize energy use. The upcycling process of plastic waste is shown in Figure 5.8. There are several benefits of upcycling such as minimizing CO_2 emission, conserving raw materials, and reducing energy and cost of production. It also supports small- and medium-scale industries to be part of plastic waste management.

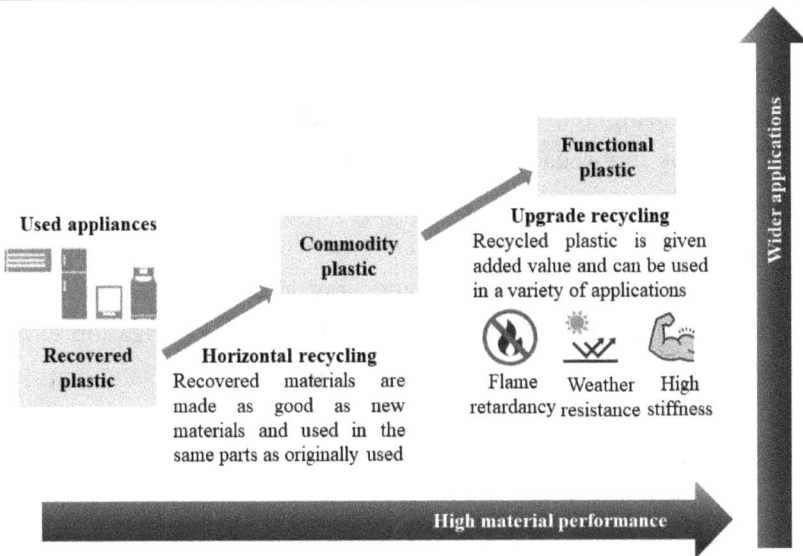

Figure 5.8   Upcycling process of waste plastics.

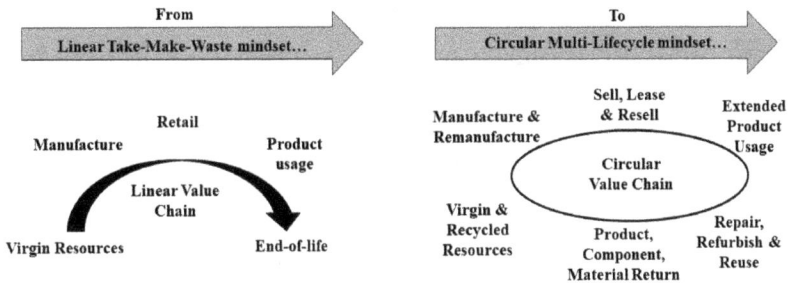

Figure 5.9   Adopting circular business models in plastic recycling. Adapted with permission.[7]

5.6 Sustainable Development Goals on Plastic Management

The UN has estimated that ~26 billion tons of plastics will be generated over the next 30 years. The environmental cost associated with the production, use, and disposal of plastics will also increase. To minimize these costs requires greater efficiency of plastic management where conventional linear economic models (i.e., manufacture-use-dispose), would change and move towards circular economic models (Figure 5.9).

It will optimize the entire life cycle of plastics using rethinking, redesign, restructuring, reuse, recycling, and recovery.

Under the umbrella of SDGs, several nations, including India and Bangladesh, have enacted bans on single-use plastic bags. It can prevent drainage systems from clogging and maintain a clean city by minimizing waste at the source. On the other hand, several countries such as Japan, China, and Singapore are advancing greenhouse gas mitigation efforts through landfill diversion and the use of intermediate waste treatment approaches. Sustainable plastic waste management can fulfill up to nine SDGs as shown in Figure 5.10.

3	Good health and well-being for people
6	Minimize pollutants entering water and avoid clogging in drainage systems
8	Waste to wealth boosts the economy and reduces the environmental cost
11	Good waste management creates healthy and sustainable cities
12	Responsible consumption of raw materials (petro-based and bio-based plastics) and decoupling economic growth from resource
13	Mitigating emissions of greenhouse gases and chlorine radicals
14	Safeguarding aquatic life of seas and oceans, strategies to combating overfishing, acidification and coastal eutrophication
15	Restore ecosystems and preserve biodiversity, reducing the inputs and impacts of waste plastic on terrestrial environment
17	International alliances on sustainable uses of plastics and minimization of marine plastic pollution

Figure 5.10 UNSDGs on plastic waste management.

Apart from this, Break Free From Plastic launched in 2016 is one of the major global movements envisaging a future free from plastic pollution and emphasizing curtailing single-use plastic from our life. Further, in March 2019 the United Nations Environment Assembly accepted a Ministerial Declaration titled *'Innovative Solutions for Environmental Challenges and Sustainable Consumption and Production'*, which promises to significantly lessen the manufacturing and use of single-use plastic products by 2030. The United Nations Environment Programme has also provided a *'National Guidance for Plastic Pollution Hotspotting and Shaping Action'* and emphasized that SDG12 and SDG14 are important for the sustainable management of plastic pollution.

A case study of Singapore indicates that 930 metric tons of plastic waste were generated in 2019. About 100% of this waste was collected by the formal sector. However, only 4% of this waste was recycled through incineration plants where they were converted into energy. The remainder were exported to other nations including Australia, China, India, Indonesia, Malaysia, South Korea, and Thailand for waste processing. In 2019, the Singapore government introduced EPR for both electronic and packaging waste, and due to this plastic producers now have a responsibility to recycle their waste. The EPR approach encourages shared responsibility and liability between all parties for plastic recycling. A new waste management plant has been set up that not only produces electricity from waste but also construction materials. In this context, Singapore Environmental Council has designed a plastic waste management infrastructure that incudes circular economy tools and EPR rules as shown in Figure 5.11. Other countries have also started good practices to recycle plastic waste. Germany has initiated a green dot in recyclable plastic products, while Taiwan's 4-in-1 recycling program is regarded as one of the best practices in the world.

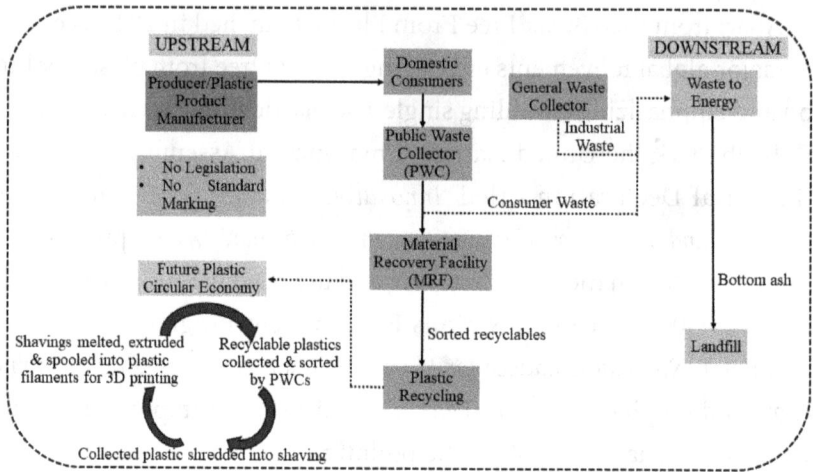

Figure 5.11   Plastic waste management infrastructure in Singapore. Adapted with permission.[8]

References

1. https://oecd-environment-focus.blog/2021/01/07/the-price-of-plastic-waste-and-solutions-to-turn-the-tide/.

2. Chamas, A., Moon, H., Zheng, J., Qiu, Y., Tabassum, T., Jang, J. H., Abu-Omar, M., Scott, S. L. and Suh, S. (2020) Degradation rates of plastics in the environment, ACS Sustainable Chem. Eng., 8, 3494–3511.

3. Alabi, O. A., Ologbonjaye, K. I., Awosolu, O. and Alalade, O. E. (2019) Public and environmental health effects of plastic wastes disposal: a review, Journal of Toxicology and Risk Assessment, 5(1), 021.

4. https://www.plasticpollutioncoalition.org/blog/2019/2/20/report-plastic-threatens-human-health-at-a-global-scale.

5. https://webstore.ansi.org/standards/astm/astmd720906.

6. Shanmugam, V., Mensah, R. A., Försth, M., Sas, G., Restás, Á., Addy, C., Xu, Q., Jiang, L., Neisiany, R. E., Singha, S., George, G., Jose, T., Berto, F., Hedenqvist, M. S., Das, O. and Ramakrishna, S. (2021) Circular economy in biocomposite development: State-of-the-art, challenges and emerging trends, Composites Part C: Open Access, 5, 100138.

7. Schwarz, A. E., Ligthart, T. N., Bizarro, D. G., Wild, P. D., Vreugdenhil, B. and vanHarmelen, T. (2021) Plastic recycling in a circular economy; determining environmental performance through an LCA matrix model approach, Waste Management, 121, 331–342.

8. http://sec.org.sg/wp-content/uploads/2019/07/DT_PlasticResourceResearch_28Aug2018-FINAL_with-Addendum-19.pdf.

CHAPTER SIX

SUSTAINABLE CONSTRUCTION AND DEMOLITION WASTE MANAGEMENT

6.1 Introduction to the Construction Industry

Waste is an inevitable by-product of any human activity regardless of a nation's economy. The construction industry is one of the critical sectors of the economy, significantly influencing a nation's social-fiscal environment and sustainability. On average, construction accounts for more than 10% of gross domestic product (GDP) worldwide (6–9% in developed countries). The global construction industry sectorized under (a) Building Construction — which includes residential and non-residential construction, (b) Infrastructure Construction — often referred to as heavy civil engineering, which includes large public works, roads, railways, dams, bridges, water or wastewater and other utility distribution, etc., and (c) Industrial Construction — includes mills and manufacturing plants, construction in offshores, mining and quarrying, oil refineries, power generation, etc., has reported an output estimated at $10.8 trillion in 2017.

The construction industry has a significant effect on the development of any nation. Being the most dynamic and responsive sector with evident output, the Gross Value Added by construction to a nation's economy was highest for China at 934.3 billion USD, followed by the United States at 839.1 billion USD (Figure 6.1). The intersectoral linkages between construction and other industries help stir the following benefits:

(a) *Attracting Foreign Investment*: A robust construction sector with accurate health and safety protocols can help to bring in foreign investment. These investments would create future capital to be invested elsewhere.

(b) *Boosting Tourism*: Improved infrastructure, including creating new public buildings or renovating heritage buildings, boosts local tourism.

(c) *Spurring Growth for Other Industries*: Infrastructure for accommodation, restaurants, built-up attractions, tours, and transport creates

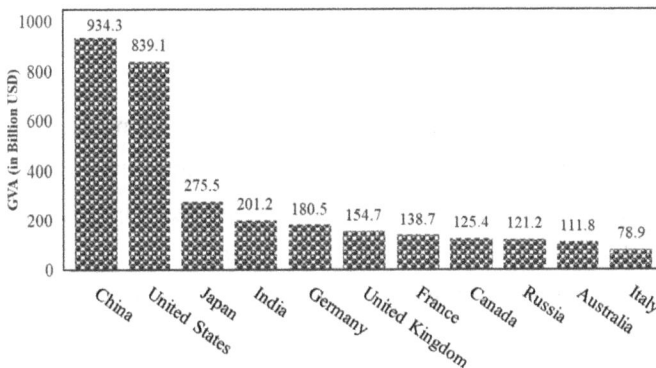

Figure 6.1   Gross Value Added (GVA) by construction for year 2019 (in billion USD): projections by region.

growth in other associated industries, primarily developed by the private sector.

(d) *Creating Jobs*: Creating new buildings requires skilled workers with knowledge of contemporary design and construction techniques. Presently 7% of the global workforce, i.e., over 273 million people, are directly or indirectly employed in construction.

Although the domino effect of the construction industry on other sectors and its contribution to the GDP of economies are indisputable, it is also responsible for the excessive consumption of natural resources, high greenhouse gas emissions, and increased production of solid wastes. Solid waste from the construction industry is referred to as construction and demolition (C&D) waste. Of the available 32% of the world's natural building materials and water resources, construction activities consume about 25% of virgin wood and 40% of raw materials drawn out from the earth's surface.

6.2 What is C&D Waste?

C&D waste is produced during new construction, repair, renovation, and demolition of small and large building apartments such as residential and non-residential buildings, and public work projects such as roads, bridges, utility plants, piers, dams, and any civil engineering structural construction projects. C&D waste is also generated due to the sudden devastation of infrastructure in natural calamities. Figure 6.2 presents a generic characterization of C&D waste based on the source of origin.

C&D waste often contains waste which is bulky in addition to having high density and occupying enormous storage space on the roadside or container facilities. These wastes are:

- Concrete/Masonry
- Wood (buildings)

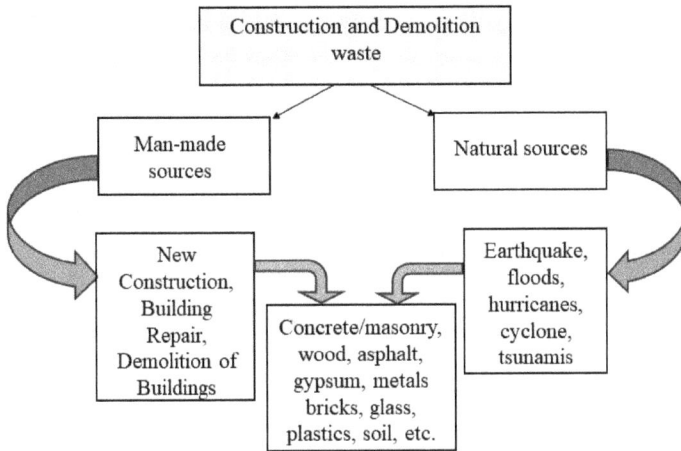

Figure 6.2 Generic characterization of C&D waste based on the source of origin.

- Asphalt (roads and roofing shingles)
- Gypsum (main component of drywall)
- Metals
- Bricks, glass
- Plastics
- Salvaged building components (doors, windows, and plumbing fixtures)
- Soil and rock from excavation
- Trees, stumps, earth, and rocks from clearing sites

They often obstruct traffic and add additional workload to the local body for dumping in a nearby landfill facility.

In 2018, the United States Environmental Protection Agency (EPA) estimated 600 million tons of C&D waste debris were generated in the country, more than twice that of Municipal Solid Waste (MSW). Thus, globally C&D waste is a concern due to the significant quantity of impermissible dumps, creating an urban menace. Unless adequately managed, C&D waste might cause adverse effects on the environment and result in air, water, and soil contamination, ultimately jeopardizing human health.

6.2.1 C&D Waste Generation in Major Countries

Buildings and civil engineering structures, when built, renovated, or demolished, generate a sizable amount of C&D waste. In general, the quantity and quality of C&D waste are greatly influenced by several factors such as:

(a) age of the structure,
(b) type of structure,
(c) construction materials used,
(d) construction technologies adopted,
(e) demolition technologies implemented,
(f) constructors' C&D waste management capabilities,
(g) population growth, etc.

Since the growth of the global construction industry is dependent on population growth, regional planning, legislation, etc., in a region, the quantity and composition of C&D waste vary between regions. The amount of C&D waste generated in some of the leading countries in recent years is presented in Figure 6.3. In China, approximately 934.3 billion tons of C&D waste were generated in 2019, from 600–800 million tons in 2011. In 2014 Beijing alone contributed around 40 million tons of C&D. In the USA, of the 534 million tons of generated C&D waste, 505.1 million tons were produced during demolition activities. Since C&D waste accounts for approximately 25–30% of solid waste in EU nations, the EU Waste Framework Directive mandated value-added recovery (or reuse and recycling) from the non-hazardous part of C&D by 2020.[1] Hong Kong successfully manages its 3 million C&D waste in 1993 to 20 million in 2013 by following good waste management practices.

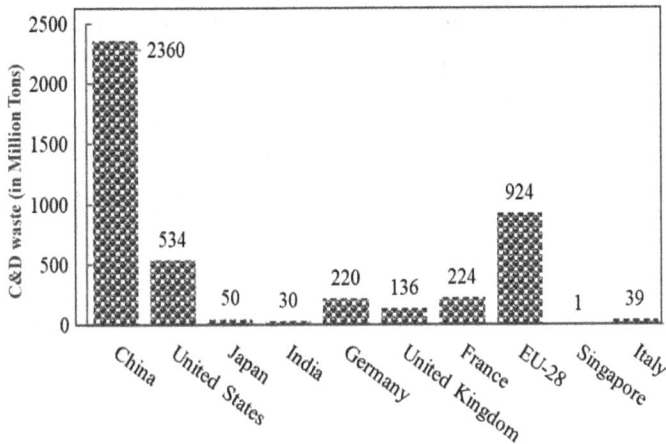

Figure 6.3 Annual C&D waste generation, in million-ton projections by region.

For India, the International Society of Waste Management reports that estimated waste generation during construction is 40–60 kg/m^2, and 40–50 kg/m^2 during renovation/repair work. While there is massive uncertainty regarding the quantum of C&D waste generation (per month or annually) in every state, it is opined that on average C&D waste accounts for about 30% of the total solid waste produced in a city in India.

6.2.2 Phases of C&D Waste Generation

The composition of C&D waste is highly dependent on the project, as described in the previous section. We can trace the origin and causes of C&D waste to improper management during the following phases:

(a) *Design phase* such as changes in design, construction detail flaws, and uncertain description of the final project.

(b) *Procurement phase* such as mistakes in shipping and ordering, faulty material delivered by the suppliers.

(c) *Materials handling phase* such as unsuitable storage and improper handling activities of the construction materials.

(d) *Transportation phase* such as damage of resource materials during transportation due to inadequate method of unloading or inferior workmanship, inappropriate resource handling from storage to the point of use.

(e) *On-site management and planning phase* such as deficiency of on-site waste management strategies and material quality control, lack of schedule for necessary amounts of materials, and no proper supervision at the site.

(f) *Residual phase* such as leftovers from application processes, wastage from packaging, etc.

6.2.3 *Composition of C&D Waste*

Most C&D waste are generally inert (Figure 6.4) and therefore will not pose an environmental hazard, unlike hazardous waste or typical MSW. For this reason, in many regions, it is not controlled in the same way

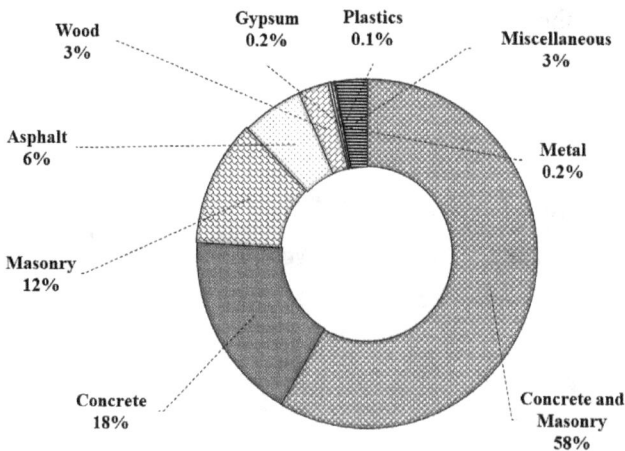

Figure 6.4   Typical categories of the materials constituting C&D waste.

as other sources of waste, with a consequent lack of data and statistics. C&D waste are usually mixed waste, in which the degree of heterogeneity depends on the category of work (demolition or construction).

In India, the demolition of old buildings creates significant waste such as soil, sand, and gravel (26%), bricks and masonry (32%), concrete (28%), metal (6%), wood (3%), and others (5%). It has been observed that in excavations, concrete, masonry, and wood constitute over 90% of all C&D waste in India. The quantum of generation of C&D waste estimates available from the Technology Information, Forecasting and Assessment Council are given below:

(a) **New Construction = 40–60 kg/m^2 of C&D waste**
(b) **Building Repair = 40–50 kg/m^2 of C&D waste**
(c) **Demolition of Buildings = 300–500 kg/m^2 of C&D waste**

In 2017, the recycling rate in France was 47%, in comparison to UK (86%) and Italy (75%). This is due to the low population density of France, with a geographically vast land area and comparatively smaller population than UK and Italy.

In Singapore, it is reported that of the 1.69 million tonnes of C&D waste generated in 2013, the rate of recycling was 99%. It is usually sorted for the recovery of materials such as ferrous metals, plastics, wood, and metal, and reprocessed into aggregates for future use in construction activities and road kerb. Sorting is done manually and by magnetic separators. Some common motivations to recycle C&D wastes include:

1. C&D wastes are one of the largest waste streams in any country.
2. Almost all waste material generated in the phases of C&D are recyclable (Table 6.1).
3. Cost of recycling is much less than landfills or discarding options.

Table 6.1　List of recyclables (under non-hazardous category) in C&D sector.

Architectural Salvage	Land Clearing	Ferrous Metals
• Doors and door frames • Wood • Panels • Plyboards, beams • Furniture and furnishings • Carpeting • Roofing • Wood, metals, plastic commercial membranes	• Soil **Asphalt** **Ceiling tiles** **Porcelain fixtures** **Aggregates** • Concrete • Brick • Concrete blocks	• Structural steel • Steel-framing members **Non-Ferrous Metals** • Wiring/conduit • Plumbing

Table 6.2　Components of C&D waste (under hazardous category).

Waste Type	Hazardous Components	Possible RCRA* Waste Codes
Construction waste	• Contaminated soil, dredge spoil, and other substances. • Asbestos-based materials are in the form of breathable fiber (toxic and carcinogenic). • Wood treated with fungicides, pesticides, etc., for carpentry and floorwork (toxic and flammable). • Gypsum-based elements (toxic and flammable).	• D001 (ignitable wreckage and debris, asphalt wastes, petroleum distillates, used oil sent for disposal, acetone, adhesives, coatings). • D006–008 (lead pipes, toxic wreckage, chromium-based paint debris). • D018 (asphalt wastes containing benzene).
Demolition waste (more homogeneous than construction waste)	• Materials containing asbestos, lead, zinc, paints, batteries, fluorescent tubes, mercury-containing wastes, air-conditioning facilities, etc.	• D023–D026 (toxic wreckage, treated wood and debris containing cresols).

*RCRA: Resource Conservation and Recovery Act.

Table 6.2 shows components of hazardous C&D waste. Wastes are defined as hazardous by EPA if they are specifically named on one of four lists of hazardous wastes (listed wastes) or if they exhibit one of four characteristics (characteristic wastes). Each type of hazardous waste is given a unique hazardous waste code using the letters D, F, K, P, or U and three digits (e.g., D001, F005, P039). Hazardous waste exhibit one or more of the following characteristics:

(a) **Ignitability:** Ignitable wastes are spontaneously combustible and have a flash point of less than 60°C (140°F). The waste code for these materials is D001.

(b) **Corrosivity:** Corrosive wastes are acids or bases that are capable of corroding metal containers. The waste code for these materials is D002. C&D debris does not typically include corrosive wastes.

(c) **Reactivity:** Reactive wastes are unstable under "normal" conditions and cause explosions, toxic fumes, gases, or vapors when mixed with water. The waste code for these materials is D003. C&D debris does not typically include reactive wastes.

(d) **Toxicity:** Toxic wastes (in C&D such as trichloroethylene, asphalt wastes, and lead pipe) are harmful or fatal when ingested or absorbed. The waste codes for these materials range from D004 to D059.

6.3 Estimation of C&D Waste Production

In recent decades, studies have been directed to the collection of data on the composition and amount of C&D waste generated in a particular geographical area or in buildings of a certain type such as dwellings, non-dwelling buildings, dwelling rehabilitation, non-dwelling rehabilitations, dwelling demolitions, and non-dwelling demolitions. This quantification is dependent on factors like (a) forecasted resource demands, (b) facilities for the management of waste, (c) targets for the reduction of waste, and (d) recovery or disposal of waste within a legislative framework.

Estimation of the amount of C&D waste generated can be determined by:

a) Per capita factors, in much the same way as MSW estimates.

b) Financial value of building permits for construction projects. The statistical data on the number, value, and area of each debris-generating activity is multiplied by the range of waste generated in each activity.

The total construction area data for the building under construction/demolition or renovation needs to be acquired from the database of the Census Bureau. Firstly, the gross floor area in the yth year (GF_y) of the building can be calculated by multiplying the covered area (CA_y) and building floor area ratio (BFR).

$$GF_y = CA_y \times BFR \tag{6.1}$$

CA_y can be calculated by deducting the urban road surface area (URS_y), and urban green coverage area (UGC_y) from the built-up area (BU_y).

$$CA_y = BU_y - URS_y - UGC_y \tag{6.2}$$

Finally, demolition area (DA_y) could be estimated after computation of additional building area for that given year (ABA_y) (if any).

$$DA_y = GF_{y-1} + ABA_y - GF_y \tag{6.3}$$

The generation of C&D waste could be calculated separately through the unit generation index $(UGI$ in kg/m$^2)$ using Table 6.3 and Eq. (6.4).

$$C \& D\ waste_{yth\ Gen} = \sum CA_y \times UGI_{CW} + DA_y UGI_{DW} \tag{6.4}$$

Table 6.3   Typical unit generation indexes (UGI) of C&D waste (kg/m²).

Typical Types of Waste	UGI-CW	UGI-DW
Metal	3.53	97.93
Concrete	18.03	674.69
Mortar	1.45	109.96
Brick/block	1.75	308.90
Ceramic	—	146.72
Timber	6.35	26.01
Others (including glass, plastics, insulation materials and mixed fragments)	3.11	54.37

Note: UGI = unit generation index; CW = construction waste; DW = demolition waste.

Another means of estimating C&D waste is through material stocks and flows using the dynamic model technique. For example, in Taiwan, Canada, the US, Europe, and Norway the quantities of C&D waste expected within a region are studied by the dynamics of collection and the flow of materials in construction activity and future building demand. Based on the schedule of work activities as per the permits issued for the construction and demolition of buildings up to, say, the year 1999, the estimate of total concrete waste generated during the period 1981 to 2011 can be determined.

6.3.1 C&D Waste Management: Circular Economy-inspired Measures

Since most of the C&D waste materials are inert in nature and bulky in characteristics, requiring more storage area, they cannot follow standard waste disposal methods and implementation of sound waste management practices becomes slightly challenging according to circular economy principles. The most relevant strategies for adopting a circular economy model in the construction and demolition sector follow the following five life cycle stages.

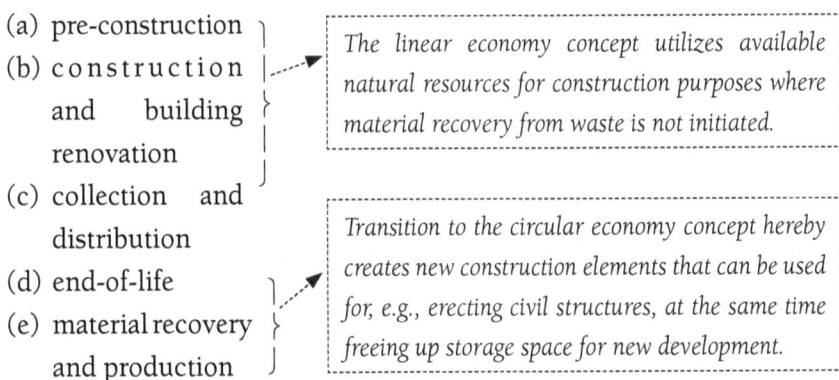

(a) pre-construction
(b) construction and building renovation

> *The linear economy concept utilizes available natural resources for construction purposes where material recovery from waste is not initiated.*

(c) collection and distribution
(d) end-of-life
(e) material recovery and production

> *Transition to the circular economy concept hereby creates new construction elements that can be used for, e.g., erecting civil structures, at the same time freeing up storage space for new development.*

As highlighted in Figure 6.5, circular economy-inspired activities can help accomplish waste policy objectives, keeping in mind the Sustainable Development Goals (SDGs) — namely, waste prevention and

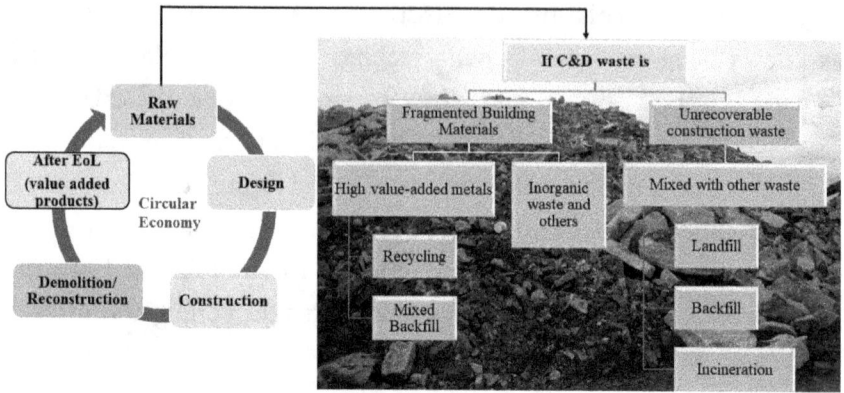

Figure 6.5   Circular economy-inspired actions for C&D waste management.

increasing both the quality and quantity of C&D waste recycling while reducing hazardous materials in the waste. Primarily, circular economy within the construction industry refers to secondary building materials continually being recycled to re-enter the economy rather than ending up as waste. C&D waste management strategies are being planned by nations to find appropriate utilization of C&D waste management-based recycled products. In 2013, EPA started including information on C&D debris generation, as this category of waste is a large part of the non-hazardous waste stream.

6.3.2 End-of-life Stage of C&D Waste

Until a few years ago, the existing C&D waste generation and supervision model was completely linear. C&D waste generated at the site was transferred to landfills, terminating its life cycle. The dumping of such wastes has created turbulence in the existing landfills dedicated to catering mostly to inert MSW. Over the last decade, a paradigm shift started, introducing the need to study C&D waste as a discrete type of waste (depending on the nature and quantity of waste generated) with the intention of recycling. However, reaching these objectives involves facing various challenges in the construction sector, from legal and administrative barriers to raising awareness in society.

6.3.3 Policies and Legislation for C&D Waste Management

Countries around the world reduce C&D waste by introducing different legislation and raising awareness. In Germany, successful plans for waste minimization and management in addition to recycling programs have brought about prospering experiences. People have recognized that construction waste is also a kind of resource that can be used for developmental activities, once treated well. In Japan, construction waste can be divided into more than 20 types, and proper recycling and recovery protocols are devised with respect to each type of waste. In Singapore, attention is more towards the concept of "green building", and prevention of the generation of construction waste is initiated from the source. Similar initiatives on C&D waste management undertaken in other countries are presented in Table 6.4.

To protect natural resources, decrease the emissions of greenhouse gases, reduce the landfill space requirement, and save money, most regulations and guidelines on C&D waste management emphasize basically:

(a) Recycling of C&D wastes
(b) R&D activities for utilization of concrete/masonry C&D wastes by establishing standards
(c) Need for popularizing the new products from C&D waste
(d) Creating quality control and certification of these new products

Table 6.4   Global initiatives on C&D waste management.

Regions	Regulations and Guidelines
EU	Management of C&D waste is steered by the EU Waste Framework Directive (2008/98/EC), which sets a target for the recycling of non-hazardous C&D waste at a minimum of 70% of its weight by 2020.
Germany Spain	National legislation governing C&D waste management requires waste generators or owners to classify, recycle and reuse C&D waste.
UK	On-site waste reduction via sorting and recycling is compulsory, in addition to recycling and reusing C&D waste.
Taiwan	Construction sites are strictly monitored since 2005 and waste generators must transport the C&D waste to treatment facilities for recycling and reuse.
Australia	C&D waste policies have been established by different state governments, promoting on-site reuse of C&D waste and implementation of sustainable building practices.
US	Federal and state highway contracts specify the use of recycled materials in the highway construction.
Singapore	The Building and Construction Authority in collaboration with the Housing and Development Board began developing a green building rating tool and using a centralized chute for recyclable wastes.
India	National Building Code (NBC 2005), Ministry of Urban Development (MoUD, 2012), Construction & Demolition Waste Management Rule 2016 work in key thrust areas regarding C&D waste reuse/recycling.

(e) New legislation and optimization for interfacing these new C&D waste-based products with traditional materials in manufacturing industries

(f) Environment protection through C&D waste utilization

6.4 Waste Management Plans and Strategies

C&D wastes are often bulked as a single waste stream, but the types of debris generated through C&D activities are vastly different. They also differ significantly in ease of separation/segregation, recovery, and recyclability. Figure 6.6 provides a framework for selecting a waste management plan.

In many countries, recycling opportunities exist for most C&D waste materials, including asphalt, concrete, drywall, metal, wood, brush, dirt, rocks, and cardboard. The main strategies include:

- To increase the collection of separately collected C&D waste
- To reduce the amounts of C&D waste generated
- To move up the waste hierarchy
- To innovate and incentivize the recycling industry

Reduce — Reuse — Recycle is the most adopted principle for managing waste in the construction industry. The philosophy of 3R also reduces the gap between the demand and supply of building materials in construction fields, especially in road construction. This further helps in the

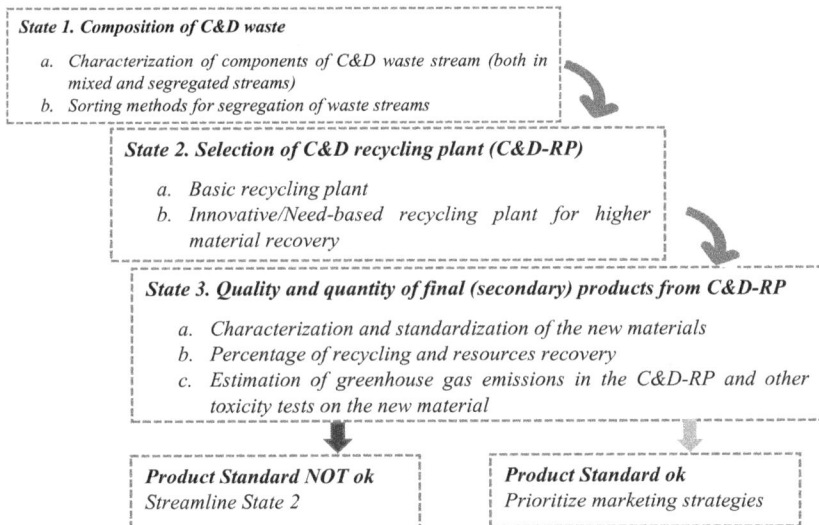

State 1. Composition of C&D waste

   a. *Characterization of components of C&D waste stream (both in mixed and segregated streams)*
   b. *Sorting methods for segregation of waste streams*

State 2. Selection of C&D recycling plant (C&D-RP)

   a. *Basic recycling plant*
   b. *Innovative/Need-based recycling plant for higher material recovery*

State 3. Quality and quantity of final (secondary) products from C&D-RP

   a. *Characterization and standardization of the new materials*
   b. *Percentage of recycling and resources recovery*
   c. *Estimation of greenhouse gas emissions in the C&D-RP and other toxicity tests on the new material*

Product Standard NOT ok
Streamline State 2

Product Standard ok
Prioritize marketing strategies

Figure 6.6   Framework for selection of the best C&D waste management strategy.

reduction of production costs, when the waste management strategies are developed from the project design phase to reduce the amount of waste materials going to landfills or unwanted material disposal sites.

6.5 Why is C&D Waste Management a Challenge?

C&D waste is a major challenge for the construction industry due to the increasing volume of waste produced and its associated environmental impacts. In the EU, C&D waste accounted for 924 million tons in 2016 (36% of the total solid waste), while in the US this amount was 534 million tons (67%),[2] and in China it was around 2.36 billion tons (30–40%).[3] As per the available statistics, global C&D waste generation will strike 2.2 billion tonnes by the year 2025. The construction sector creates serious environmental problems, especially during the operation and end-of-life stages of the construction process. The devastating environmental effects include resource depletion, land degradation, landfill depletion, carbon and greenhouse gas emissions, water pollution, and high energy consumption. Hence, it has become a priority for sustainable development programs worldwide. Even though there is growing interest in employing recovery practices, i.e., reuse and recycling C&D waste, in most cases the process of waste management is inefficient, resulting in a substantial amount being disposed of in landfills or illegally dumped without proper environmental protection measures. The World Economic Forum reported in 2016 that only 20–30% of C&D waste is recovered globally.[4] As shown in Figure 6.7, the lowest recovery rate is in China at less than 5%. Under Waste Directive 2008/98/EC, an average of around 70% recovery and recycling was targeted for EU nations by 2020, but only 46% was achieved. All this means the current waste management system needs to be improved for many nations.

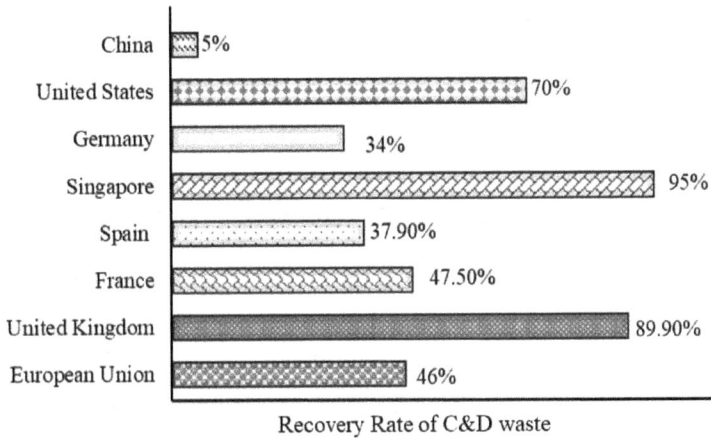

China   5%
United States   70%
Germany   34%
Singapore   95%
Spain   37.90%
France   47.50%
United Kingdom   89.90%
European Union   46%

Recovery Rate of C&D waste

Figure 6.7   Global trends of recovery of value-added materials from C&D waste.

6.5.1 *Current Barriers to Sustainable C&D Waste Management*

The current obstacles to sustainable C&D waste management have been identified as follows:

1. Non-confidence of stakeholders in the use of products derived from waste, and low public participation and awareness.
2. The non-availability of stringent legislation for recycling of C&D waste.
3. The absence of on-site sorting as well as selective demolition practices reduces the prospective quality of recycled aggregates.
4. Limited knowledge of the technical aspects of recycled aggregates reduces the use of C&D recycled materials.
5. Recycled aggregate quality does not meet expectations due to mixups with other materials. They are also not always competitive compared to aggregates from virgin materials, due to the lack of taxes on mining activities.
6. The difference in price between recycled building materials and virgin materials is negligible.

7. Overreliance on the private sector for waste management.

8. Limited treatment options.

6.6 Looking to the Future: How can the Construction Industry Contribute to Its Waste Management Plan?

An effective construction and demolition waste management policy service is substantially dependent on long-term sustainability and cost factor to combat the industry's waste problems. Listed in Table 6.5 are the following components that should be considered by waste generators in the preparation of an efficient waste management plan.

Table 6.5   Stepwise guidance in the preparation of a construction and demolition waste management plan.

Elements of Sustainable Construction Waste Management	Action	Effect on Societal Development and Environmental Protection
Construction and demolition waste management plan	• Calculate amount, composition, or quantity of waste that will be produced during a construction project. • Detail storage, transport, handling, transfer, disposal, and processing of waste. • Assess for hazards to the environment or human health associated with the waste.	• Ensure businesses associated with the generated waste are compliant with the five-tier waste hierarchy, i.e., prevention of waste altogether, followed by reduce, reuse, recycle, and dispose of.
Material resource efficiency model	• Ensure a more transparent construction supply model chain, not entirely focusing on a project-by-project basis of procurement.	• Involvement of all stakeholders (viz. procurement officers, construction project managers, contractors, and clients) in the overall procurement process.

Table 6.5   (*Continued*)

Elements of Sustainable Construction Waste Management	Action	Effect on Societal Development and Environmental Protection
Waste classification and disposal	• Classify according to its characteristics into reusable, recyclable, unwanted and hazardous categories.	
Pre-demolition and renovation audit	• Provide detailed information on materials to be reclaimed and recycled in advance for any project works.	• Reduction in overall cost and environmental impact of waste disposal.
Regular monitoring during project tenure	• Provide actual waste data versus forecasted waste data, and test the efficiency of the model.	• Ensure objectives outlined in the plan are met.
Public participation and awareness	• Encourage public education and community participation in issues related to C&D waste. • Promote the use of recycled products in new constructions by state administrations. • Impose a legislative framework to use a minimum percentage of recycled material in public works projects.	• Increased usage of C&W recycled material.

6.6.1 *Case Study of Singapore's C&D Waste Management and Climate Benefits*

Singapore contributes in different percentages to the total generated waste globally, highest being construction debris (18.40%), followed by ferrous metal (17.57%), waste from paper and cardboard (15.54%), and food waste (10.24%). The National Environment Agency of Singapore in 2020 reported that the recycling rate of C&D waste is at 99%.

Construction and maintenance of the built environment consume almost half of all materials entering the global economy, generating 20%

of greenhouse gas emissions. In this regard, the Green Building Leadership under the aegis of the Building and Construction Authority in Singapore has actively endorsed the use of green building technologies and designs in building construction throughout the nation. This leads to lower demolition rates and reduced generation of C&D waste. Further CO_2 savings from better C&D waste management by introducing circular economy action is another benefit.

References

1. https://single-market-economy.ec.europa.eu/news/eu-construction-and-demolition-waste-protocol-2018-09-18_en.
2. https://www.epa.gov/sites/production/files/2016-11/documents/2014_smmfactsheet_508.pdf.
3. Huang, B., Wang, X., Kua, H., Geng, Y., Bleischwitz, R. and Ren, J. (2018) Construction and demolition waste management in China through the 3R principle, Resources, Conservation and Recycling, 129, 36–44.
4. https://www3.weforum.org/docs/WEF_Shaping_the_Future_of_Construction_full_report__.pdf.

CHAPTER SEVEN
SUSTAINABLE MUNICIPAL SOLID WASTE MANAGEMENT

7.1 Introduction to Waste

Waste is an inevitable by-product of human activity. In nature, there is no waste, because nature can "recycle" all elements in the ecosystem. However, in the human environment, waste is a product or substance which is no longer suited for its intended use.

With an increasing population, waste management has emerged as a challenging goal for municipalities and urban local agencies because of aesthetic and

- "Wastes are substances or objects which are disposed of or are intended to be disposed of or are required to be disposed of by the provisions of national law." — *Article 5 of The Basel Convention*
- "Wastes are substances or objects, other than radioactive materials covered by other international agreements, which: i) are disposed of or are being recovered, or ii) are intended to be disposed of or recovered; or iii) are required, by the provisions of national law, to be disposed of or recovered." — *Organization for Economic Co-operation and Development (OECD)*
- "Waste means any substance or object which the holder discards or intends or is required to discard." — *Article 3 of the Waste Framework Directive of the European Union*

environmental concerns. The key trends driving waste management include:

(a) increase in the daily generated volume of waste in urban societies,
(b) change in the quality or characteristics of the generated waste,
(c) identifying the most suitable disposal method for the collected waste.

Municipalities and urban local agencies have, time and again, identified solid waste as a significant problem requiring drastic measures. The changes in the management of solid wastes generated in urban societies over the past few decades were dedicated to reducing hazardous and potential exposures of waste to both human health and the environment. Solid waste management and its minimization is not an isolated phenomenon that can be easily grouped and solved with innovative engineering or technology. It can be termed as an urban problem that is thoroughly related to changes in urban lifestyles, resource consumption patterns, take-home pay levels, and other cultural and socioeconomic issues. Thus, decision making at household, neighborhood, city, and national levels collectively will help ensure a long-term solution to this "millennial problem".

7.1.1 Sources of Municipal Solid Wastes

Municipal Solid Waste (MSW) includes wastes generated from residences, industries, institutes, commercial establishments, construction and demolition, municipal services, etc. It is a biomass waste type consisting of food wastes, textiles, glass, metals, wood, plastics, yard trimming, rubber, and paper. Figure 7.1 shows the waste-generating sectors and the type of solid waste from each source in detail.

Commonly known as garbage or trash, MSW consists of all daily use items thrown away after use, such as food scraps, bottles, clothing,

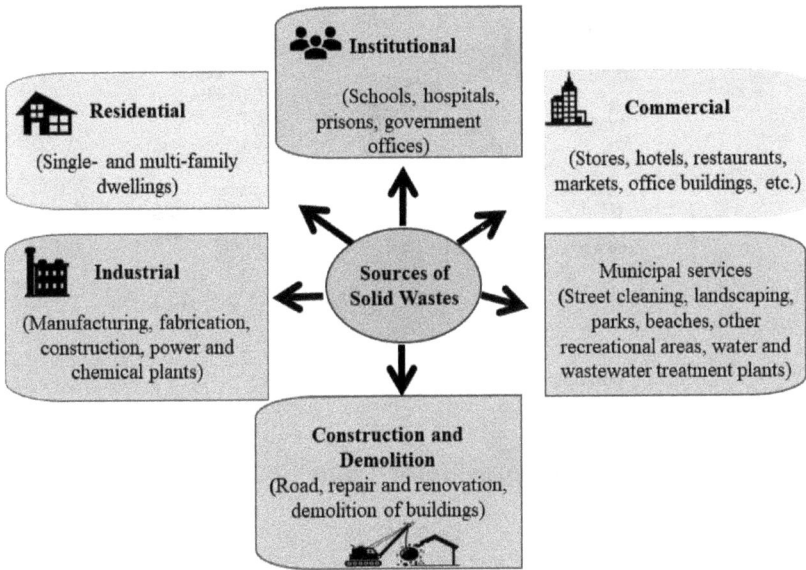

Figure 7.1   Waste-generating sectors.

batteries, paints, appliances, newspapers, furniture, grass clippings, and packaging. Thus, it includes biodegradable and non-biodegradable waste in nature and consists of electrical-electronic waste, mixed waste such as clothing, hazardous waste (paints, sprays, and chemicals), and medical waste.

As evident from Figure 7.2, MSW is the most critical type of solid waste because of its heterogeneous nature. The quantity of waste produced is highly influenced by population growth and consumption. The climatic condition, culture, and socioeconomic level of a community are some of the other factors responsible for the variation in MSW composition across different places. Figure 7.3 maps out the global trends of solid waste management by region. The trends depict that the worldwide increase in waste generation annually is expected to be significantly higher for South and East Asian countries by 2050.

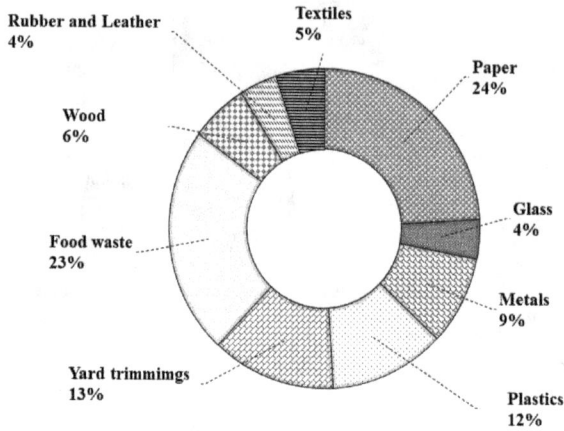

Figure 7.2 Typical generation of waste: by waste category.

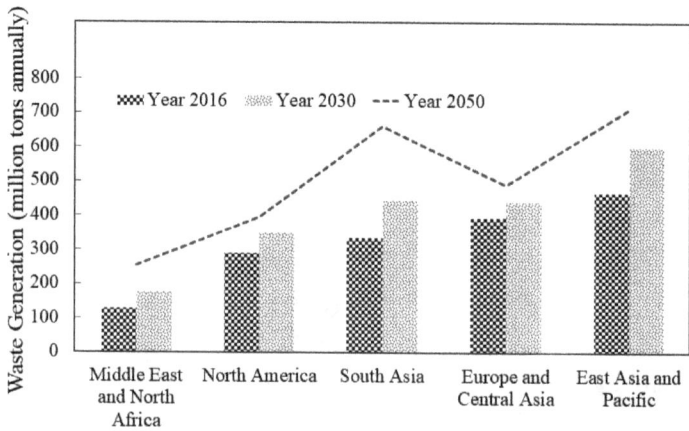

Figure 7.3 Projected annual waste generation (in million tons), by region. Reproduced with permission.[2]

Developed economies, such as the United States, Europe, and the United Kingdom, generally produce substantial amounts of per capita solid waste compared to developing and underdeveloped economies, as depicted in Figure 7.4.

The OECD countries generate 572 MT of solid water per year, with a per capita value ranging between 1.1 kg and 3.7 kg per person per day with an average of 2.2 kg per person per day. The amount of waste

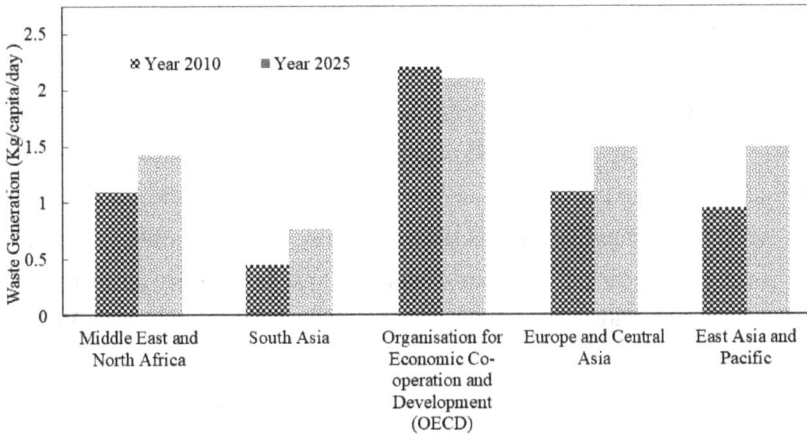

Figure 7.4   Waste generation per capita (kg/capita/day), by region. Reproduced with permission.[2]

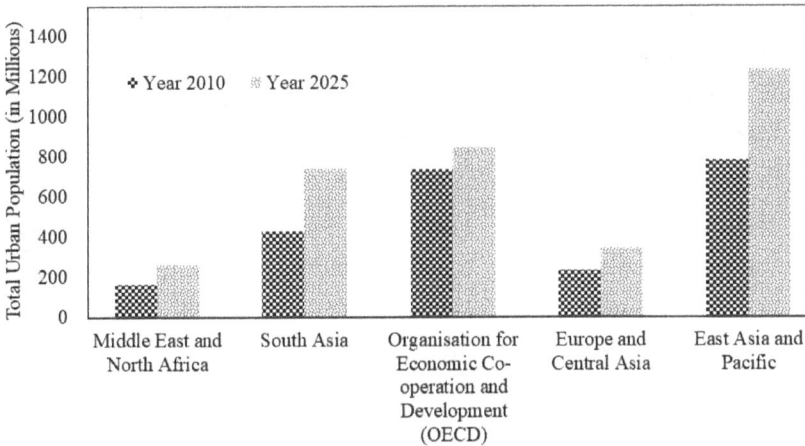

Figure 7.5   Total urban population (in millions), by region. Reproduced with permission.[2]

generated is 0.12 kg to 5.1 kg per person per day, with an average of 0.45 kg/capita/day for Southeast Asia. Comparing the OECD countries and countries of Southeast Asia demonstrates that the greater the income level and rate of urbanization (Figure 7.5), the greater the amount of solid waste produced.

In developing countries, solving the problem of MSW generation is more challenging as compared to developed economies due to (a) the low socioeconomic level of most of the population, (b) the lack of awareness of MSW issues, and (c) shortage of a suitable technology platform needed to face the problem. The World Health Organization pointed out that solid waste management is becoming a significant public health and environmental concern in urban areas of many developing countries due to the non-availability of an organized means of controlling solid waste. This necessitates the need for studying the second key objective of MSW, i.e., understanding the nature or characteristics of the generated waste to identify the most suitable method for control or disposal.

7.1.2 Characteristics of Municipal Solid Waste

MSW is a highly heterogeneous material that varies widely from place to place and time. The characteristics of MSW are influenced by a multitude of factors, *viz.* weather, geographic location, food habit, income level, economy, etc., and also change with age and decomposition.

Because of its heterogeneous nature, the determination of waste composition is cumbersome. The use of statistical procedures is problematic, and usually strategies based on random sampling techniques are utilized to determine the accurate and precise (physical and chemical) characteristics of MSW, as pointed out in Figure 7.6. The range and

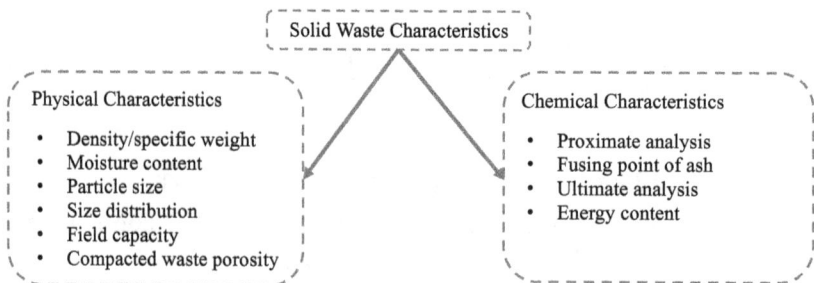

Figure 7.6   Factors affecting solid waste characteristics.

Table 7.1   The range and typical values of various attributes of MSW.

Component*	Density Range (kg/m³)	Typical Density	Moisture Content Range	Typical Moisture Content	Volatiles Content	Carbon Content	Ash	Energy Content (MJ/kg)
	(kg/m³)			(%)				(MJ/kg)
Food waste	130–480	290	50–80	70	22	4	5	3.48–6.97
Paper	40–130	89	4–10	6	75	8	5	11.63–18.60
Plastics	40–130	64	1–4	2	95	2	2	27.91–37.21
Yard waste	65–225	100	30–80	60	42	7	0.5	2.32–8.60
Glass**	160–480	194	1–4	2	—	—	—	0.11–0.23
Tin**	50–160	89	1–3	1	—	—	—	0.23–1.16

*Unlike that of Western countries, the solid wastes of Asian cities are often comprised of 70–80% organic matter.
** "—" means data not determined.

typical values of physical attributes for various components of solid waste are presented in Table 7.1.

Elucidating the physical and chemical characteristics are the first steps in planning, designing, operating, and upgrading the existing solid waste management systems. It also assists in deciding the appropriate option, i.e., incineration, composting, or landfilling for MSW disposal.

7.2 Important Landmark Policies and Initiatives on Solid Waste Management

To address the huge volumes of municipal and industrial solid waste generated nationwide, regulations are issued by agencies, such as the United States Environment Protection Agency, that translate the general mandate of a statute into a set of requirements in an open and public manner. The key element of any nation's environmental policy is to manage waste in an environmentally sound way keeping in view the following:

- To conserve energy and natural resources
- To reduce the amount of waste generated

- To protect human health and the environment from the potential hazards of waste disposal
- Prevent future problems caused by irresponsible waste management.

Solid waste legislation in the United States has been constantly strengthened and improved by introducing amendments to the four significant laws introduced under the federal government, as highlighted in Table 7.2. These federal guidelines have been instrumental in providing state, city, and some local governments regulatory responsibility for ensuring proper management of wastes generated from each source in their region.

7.3 Why is Solid Waste Management a Challenge?

The increase in population, rapid urbanization, flourishing economy, and rise in community living standards are the prime reasons for the accelerated generation rates of MSW in developing countries. The management of this increasing waste is becoming a major public health and environmental concern in urban areas of many developing countries. According to the World Bank, 2.01 billion tons of MSW are generated annually worldwide, with only 67% managed in an environmentally safe manner.[1] Until recently, the common man's perception of solid waste management was guided by "not-in-my-backyard" syndrome and dumping the waste (without segregation at the source) in the dump-yard, to be taken care of by local municipal bodies. This led to a huge pile-up of poorly managed waste resulting primarily in the contamination of (a) surface and groundwater sources, (b) soil contamination, and (c) air pollution from the burning of waste.

The other catastrophic effects on human health due to poor waste management are:

1. Transmitting diseases like diarrhea, dysentery, cholera, typhoid, salmonellosis, plague, and hepatitis via the breeding of vectors at the waste disposal sites.

Table 7.2 Landmark policies and initiatives on solid waste management in the United States and Singapore.

Year	Legislation	Regulations and Guidelines
1965	US Solid Waste Legislation: Solid Waste Disposal Act	1. For better management of solid wastes by enforcing regulations for collection, transportation, and disposal. 2. For providing financial support to different states in developing solid waste management plans. 3. For providing support to research and development on wastes.
1970	Resource Recovery Act	1. For emphasizing solid waste disposal to recycling and energy recovery. 2. For reviewing the nature, characteristics and disposal of hazardous waste. 3. For studying the early stages of solid and hazardous waste management schemes. The United States Environment Protection Agency was formed in the interim.
1976	Resource Conservation and Recovery Act	1. For control of hazardous waste storage, treatment and disposal under Subtitle C. 2. For management of non-hazardous solid waste, such as household garbage and non-hazardous industrial solid waste under Subtitle D. 3. For establishing requirements for the design and operation of underground storage tanks aimed at preventing accidental spills of toxic substances and petroleum products under Subtitle I.
1984	The Hazardous and Solid Waste Amendment	1. Revision of criteria for landfills receiving hazardous household waste or small quantities of industrial hazardous waste. 2. Treatment of contaminated surface water around landfills. 3. Regulations for groundwater protection by providing a bottom liner and leachate collection system in landfills. 4. Standardization of landfill design criteria. Methods for disposal of wastewater sewage sludge at landfills as per Clean Water Act 1970.
2019	Singapore's Resource Sustainability Act	1. Impose obligations relating to the collection and treatment of electrical and electronic waste and food waste; to require reporting of packaging imported into or used in Singapore. 2. To regulate persons operating producer responsibility schemes, and to promote resource sustainability.

2. Respiratory and visibility problems through airborne particles from burning of waste and other chronic respiratory diseases due to inhalation of emitted gases at the waste disposal sites.
3. Decreasing birth weight for people born and living within 2 km of the sites due to hazardous and non-hazardous waste.
4. Increasing physical and chemical hazards associated with direct contact with sharps and/or infected sharps, using certain toxic chemicals in unsegregated waste.
5. Increasing fire and explosion hazards due to certain solvents and emitted greenhouse gases (predominantly methane and carbon dioxide) in the transformation processes.
6. Affecting the economic development of a region due to diminished tourism.

7.4 Waste Management and Sustainability

Since prehistoric times, humans have always preferred to live in clean, hygienic conditions while keeping away the trash with the least possible effort. At the dawn of civilization, the amount of waste generated was insignificant due to low population density and abundance in the availability of natural resources. Throughout most of history, disposal methods were very crude, involving wastes thrown onto unpaved streets and roadways, or large open pits located just outside the cities. As the towns grew and populations increased, efforts were made to transport waste farther out from the cities. Another crude way for waste disposal was to dump waste in rivers flowing downstream from the city. Since the beginning of the Industrial Revolution, a technological approach to solid waste management began to develop. The change of focus from public cleansing to modern waste management was primarily driven by industrialization, which dramatically changed the types and composition of waste and made waste management complicated and costly in urban areas. The technological advances included the development of garbage bins, garbage vehicles, compaction trucks, pneumatic collection systems, grinders, incinerators for the volume reduction of waste, etc. Sanitary landfills were developed to replace open dumping and reduce the reliance on waste incineration.

With the world becoming increasingly urbanized, challenges and demands associated with providing infrastructural needs and services such as waste collection and disposal are also growing. However, of late, the economic and societal modus operandi of nation development was to utilize/extract the primary resources, transform/produce them into value-added products, and discard them when the product is no longer of use, i.e., take-make-discard. The production and consumption models were primarily linear, which seemed to be highly inefficient and unsustainable with time. In recent years, there has been some controversy over the role of technology in meeting sustainable development goals. Over the last century, economic and social progress in the urban cities assisted by the linear economy model has been accompanied by environmental degradation that is endangering the natural ecosystems on which our future development depends. This necessitated shifting from *linear to circular economy*, which mimics nature by converting waste into other value-added resources after the end of life (Figure 7.7). A circular economy promotes the reduction of consumption, considering a product's useful life and, at the end of that, its reuse or recycling. This concept of sustainability and the circular economy has progressed the MSW management system from basic disposal to recycling and resource recovery.

Currently, 2.01 billion tonnes of MSW is generated globally, out of which only 67% is collected by municipalities.[1] This annual global waste is expected to increase to 3.4 billion tons by 2050. Presently, from the total MSW that is collected by the municipalities, about 70% ends up in

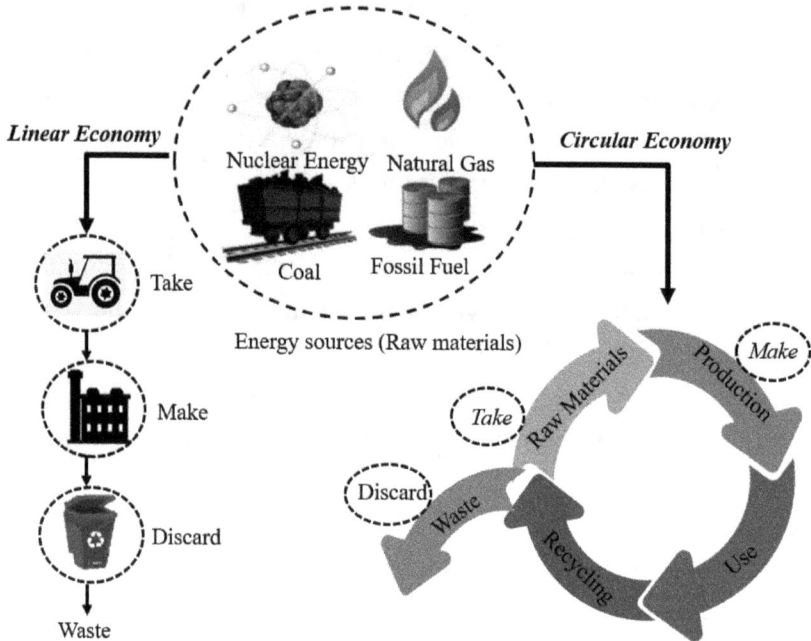

Figure 7.7   Transition from linear to circular economy.

landfills and dumpsites, 19% is recycled, and only 11% is used for energy recovery. Municipalities often face problems in assorting MSW mainly due to a lack of organization, complexity, and non-availability of financial resources.

7.5 Modern Municipal Solid Waste Management

A typical waste management scheme is depicted in Figure 7.8. The waste management sector went by the generate-collect-dispose mechanism and, since the early 1970s, adopted the 'three Rs' — reduce, reuse, recycle. The concept of sustainability and the circular economy has initiated the fourth R — recovery — in the waste management hierarchy, which led to the evolution of the Integrated Solid Waste Management System.

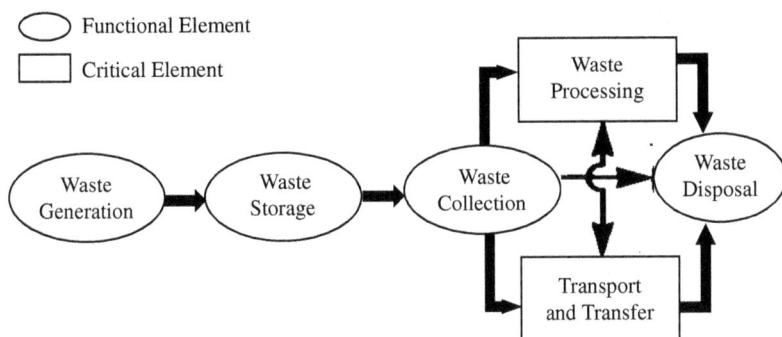

Figure 7.8   Factors affecting waste management systems.

7.5.1 Integrated Solid Waste Management System

The Integrated Solid Waste Management (ISWM) system is based on the waste management hierarchy (Figure 7.9) to reduce the amount of waste being disposed of while maximizing resource recovery and efficiency. The ISWM concept, as described, is closely linked to the 3R approach, i.e., reduce, reuse, and recycle, with more emphasis on the fourth R, i.e., recovery.

The following strategies are embraced within the ISWM hierarchy:

At-source reduction and reuse:	Waste minimization and sustainable use of products at various stages of product design, production, packaging, use, and reuse. The most preferred option for waste management.
Waste recycling:	Processing non-biodegradable waste to recover commercially valuable materials, e.g., plastic, paper, metal, glass. Next preferred alternative option for waste management.
Waste to composting:	Handling organic fraction of waste to recover compost (e.g., windrow composting, in-vessel composting, vermicomposting) and improve soil health and agricultural production.
Waste to energy:	Where material recovery from waste is not possible, recovering energy before final disposal of waste via process like bio-methanation, pyrolysis, gasification, waste incineration, production of refuse-derived fuel.
Waste disposal (landfills):	Residual waste at the end of the hierarchy, which ideally comprises inert waste, are to be disposed in sanitary lined landfills.

Greenhouse gases (GHGs) generated by the waste industry are uncontrolled and fugitive emissions, with methane comprising more than 90%, mostly in waste management landfills. ISWM reduces these methane-rich GHG emissions resulting from MSW. This is because waste prevention and recycling (including composting) divert biogenic/organic waste from reaching the landfills, thereby reducing the methane released from decomposing the organic/biogenic component of the trash at the disposal facilities.

7.5.2 Decentralized Waste Management Systems

Decentralized or community-level waste management systems (DWMS) reduce the burden of handling large volumes of MSW at a centralized location. The management of organic waste through home- and community-level composting systems are examples of DWMS. The following are advantages of a DWMS:

(a) Allows for a lower level of mechanization than the centralized solutions.
(b) Provides job opportunities for informal workers and small entrepreneurs.
(c) Options can be modified for the local waste stream, climate, social, and economic conditions.
(d) Reduces the cost incurred for the collection, transportation, and disposal of waste.

The disadvantages associated with a DWMS include:

(a) Scarcity of land within the urban locality.
(b) Difficulty in maintaining scientific and hygienic conditions owing to lack of space and trained workers.
(c) Uncertain quality of end products.
(d) Trouble in ensuring the economic viability of the system.

7.5.3 *Extended Producer Responsibility*

Extended producer responsibility (EPR) is a policy-driven approach wherein a producer is responsible for the post-consumer stage of a product. EPR programs are commonly made mandatory through legislation but can also be implemented voluntarily in retail take-back programs. EPR initiatives are typically for

(a) defined tasks of separate collection (e.g., for battery waste, e-waste, or hazardous waste),
(b) reuse (e.g., disposal-refund systems for bottles and papers),
(c) recycling (e.g., for used cars and glass containers), and
(d) storage and treatment (e.g., for batteries).

National and state-level involvement are necessary to ensure that EPR initiatives are successfully implemented.

7.6 Looking to the Future: How can Communities Contribute to the Current Municipal Solid Waste Management Plan?

The municipal authorities have to operate efficiently to keep the cities and towns clean and dispose of the MSW in an environmentally acceptable manner. An effective MSW management service is substantially dependent on community behavior and practices. Segregation of waste at source, delivering waste to doorstep collectors, participating in waste recycling, avoiding littering, buyback programs and, most importantly, exploring options for waste minimization are highly dependent on active and appropriate public involvement and support. Stepwise supervision for local authorities in the preparation of efficient MSW management plans has been listed in Table 7.3.

Table 7.3 Stepwise guidance to local authorities in the preparation of MSW management plans.

Details of MSW Management	Community Level	Municipality Level
Consolidated details of existing data on both community and municipality levels	• Current demographic data of the population • Inventory of the existing bins, vehicles, and available land for DSWM facilities • Quantity of waste generated at the source • Community engagement and outreach services	• Expected growth in population as per census data including the floating population • Inventory of equipment, bins, vehicles, and available land for ISWM facilities • Physical and chemical composition of waste
Source storage and segregation	• Details of waste storage at source and source segregation	• Details of waste storage at source in other sectors except residential, and source segregation
Collection system (in practice)	• Door-to-door collection with frequency of collection • Collection from community bins • Frequency of street sweeping	• Collection from storage in covered street bins, containers, masonry, concrete bins, open waste storage sites, or any other method • No secondary storage (direct transportation of waste)
Transportation system	• Quantity and percentage of waste transported each day in covered and open vehicles with frequency of transportation • Type and number of vehicles used	
Processing of waste	• Quantity and percentage of waste processed • Technology adopted • Revenue from the processing facility • Waste sent to disposal site	• Quantity and percentage of waste processed • Technology adopted • Revenue from the processing facility • Waste sent to disposal site
Disposal of waste	• Land availability for DSWM like composting, incineration, etc., within the community	• Identifying location of existing dump sites and capacity of existing landfill • Volume of the current cell and expected life, quantity of waste deposited annually at the landfill • Land availability for MSWM as per city development plan

References

1. http://www.worldbank.org/en/topic/urbandevelopment/brief/solid-waste-management.
2. https://stats.oecd.org/index.aspx?DataSetCode=MUNW.

CHAPTER EIGHT
SUSTAINABLE MANAGEMENT OF WATER AND WASTEWATER

8.1 Introduction to Sources of Water

Water is an essential resource required to sustain life on Earth. Of the 71% of the Earth's surface that is covered with water, about 96.5% is in the ocean, which is not conducive for direct human consumption. The freshwater resources like (a) surface water sources including lakes, ponds, reservoirs, rivers, streams, and wetlands, and (b) groundwater including natural springs, wells, boreholes, and infiltration galleries are the primary sources of water for everyday use. Surface waters hold only a small volume of the Earth's total freshwater (0.3%) supply but help in the execution of many functions in the environment, and provide people with the prime source of drinking water, energy, and recreation, as well as a means of irrigation and transport. Groundwater sources are underneath the ground surface. Unlike surface water, groundwater quality is typically constant over time; however, changes in hydrogeological conditions can lead to differences in water quality over a relatively short distance.[1]

8.1.1 *Role of Natural Water Recycling in the Hydrologic Cycle*

Water is in constant circulation and transferred between the land surface, ocean, and the atmosphere in a natural process called the "*Hydrologic Cycle*".

As evident from the cycle in Figure 8.1, water evaporates from surface storage reservoirs like rivers, lakes, oceans and land surfaces into the atmosphere; it is later advected in the form of water vapor. Eventually, condensation of this water vapor leads to precipitation in the form of rain, snow, or hail back to the Earth's surface. A significant portion of the precipitation is collected as surface runoff in surface storage reservoirs/sources. The water that infiltrates into the ground surface percolates into deeper zones to become a part of groundwater storage sources. A small part of the precipitation is transpired by vegetation, while evaporation also takes place from soil. In a final step, water re-enters the ocean and eventually evaporates, thereby completing the hydrologic cycle.

The sources and properties of natural waters in the hydrologic cycle are tabulated in Table 8.1.

It is worth mentioning that although the water cycle continuously returns water to the Earth surface, it is not always returned in the same

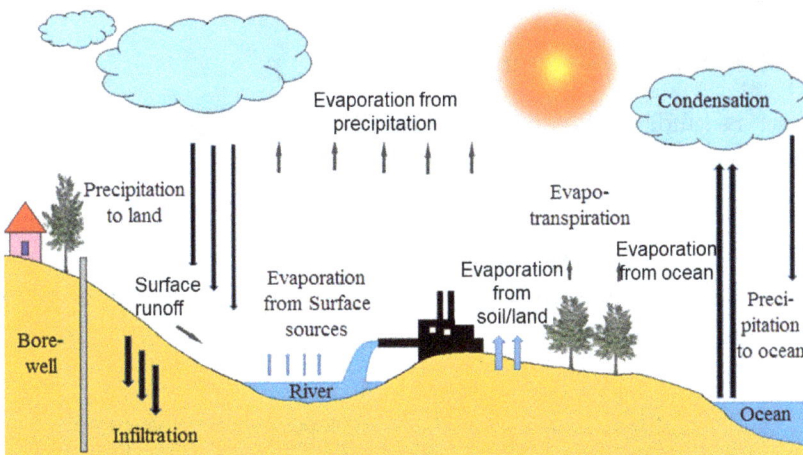

Figure 8.1   Circulation of water on the Earth's surface.

Table 8.1　The sources and properties of natural waters in the hydrologic cycle.

Source of Water	Components
Precipitation	Gases, particulate matter, salt nuclei (principally chlorides), radioactive fallout
Surface runoff	Organic matter, particulate matter (silt, clay), natural and synthetic fertilizer, nitrates, phosphates, soil bacteria
Surface water sources (rivers, lakes, etc.)	Organic matter, particulate matter (silt, clay), wastewater, algae and similar organisms
Groundwater	Carbonates, sulfates, chlorides, etc., depending on the aquifer characteristics
Seawater	Salts

quantity or quality to the same place. Surface water and groundwater are both important sources of water supply for community needs. Groundwater is a common source for single homes and small towns, whereas rivers and lakes are the usual sources for large cities with municipality services. Since surface water is more easily accessible than groundwater, it has been relied on for many human uses for ages. But wide fluctuations in the quantity of surface water due to seasonal changes have made groundwater sources a more reliable option than surface water sources.

8.2　Need for Water Sustainability

Water is fundamental to the three dimensions of sustainable development, i.e., society, economic growth, and environment. Yet water scarcity is a significant and growing problem. The demand for water in municipal needs has skyrocketed in the past few decades in both developing and developed nations. An increasing population means new demands for basic needs that contribute to additional pressure on freshwater resources.

By 2025, it is estimated that 1.8 billion people will live in countries with absolute water scarcity, and two-thirds of the world's population will be under water-stressed conditions. Annually, if a region extracts water supplies below 1,700 m^3 per person, it is defined to be experiencing water stress. Annual water supplies dropping below 1,000 m^3 per person, the region faces water scarcity, and below 500 m^3 is absolute scarcity. In water-stressed regions, the rate of over-pumping of underground aquifers is much faster than the rate of replenishment. This results in a precipitous lowering of the groundwater levels.

The factors that impact water scarcity in nations are:

a) *Population-water balance*
b) *Land use and land cover changes*
c) *Impact of climate change — alterations in rainfall pattern*
d) *Environmental degradation*
e) *Impact of pollution*

A larger population requires greater agricultural commodities, which leads to an increase in water demand. Presently, about 70% of available freshwater is used for agricultural practices; however, due to inefficient agricultural methods and faulty irrigation systems, wastage of water amounts to 60%. Higher freshwater demands in agricultural farm fields exposed to the use of pesticides and fertilizers also contribute to environmental pollution of surface water sources.

Population explosion and rapid socioeconomic development also impact the proper land use planning of a region. The accelerated rate of urbanization results in an increase in the amount of impervious area thereby reducing the rate of infiltration, and thus affecting groundwater recharge and storage. The change in land use/land cover patterns, due to human activities, has exerted small- to large-scale variations on the hydrological cycle as well. Changing the hydrological environment changes the temperatures and weather conditions of a region affecting the availability and distribution of rainfall, snowmelt, surface water, and groundwater flows, in addition to deteriorating the quality of the water.

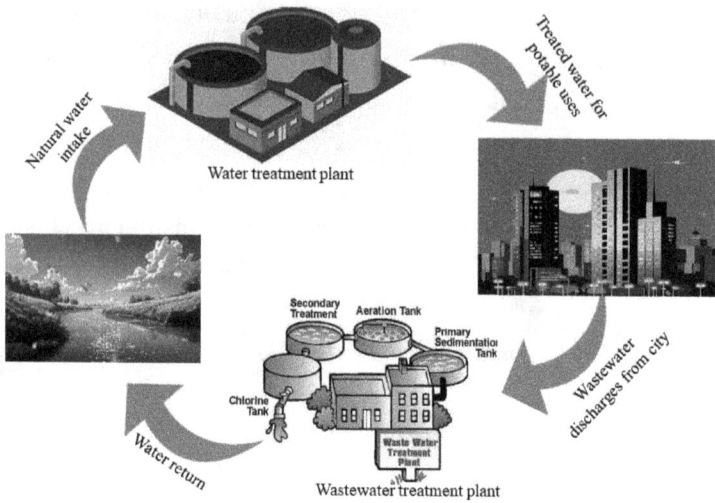

Figure 8.2   The urban water cycle: an engineered approach for sustaining modern cities.

Such changes in water availability also create a significant impact in sectors of health and food security, having the potential to trigger refugee crises and political instability.

Thus, to address the challenges posed by urban societies, *"Urban Water Cycle"*, a water management approach involving efficient use of available resources, protection of degraded ecosystems, treatment of wastewater, etc., is introduced in Figure 8.2.[2]

8.2.1 *Causes of Water Scarcity: Water Pollution*

The accelerated degradation of ecosystems due to pollution is one of the most significant causes of water scarcity across the world. According to the World Health Organization (WHO), "Safe water supplies, hygienic sanitation, and good water management are fundamental to global health."

Figure 8.3 highlights the pollution of water either by point sources or non-point sources, which threatens fundamental global health initiatives. Point source pollution is generally confined, discernible, and

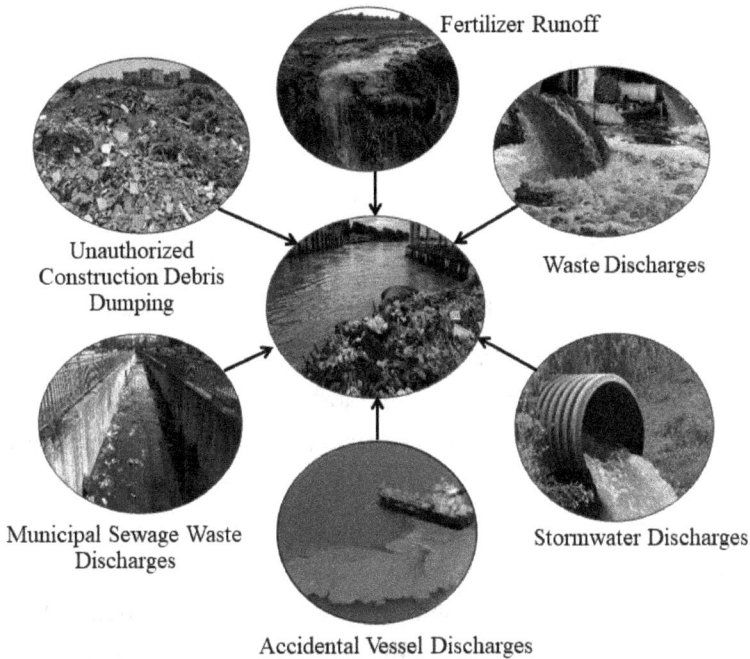

Figure 8.3   Reasons for water pollution due to point and non-point sources.

discrete, e.g., water treatment plants, landfills, oil spills, and sewage discharge. In contrast, non-point source pollution is when a body of water is polluted from multiple sources, e.g., runoff water from an agricultural field containing herbicides, insecticides, fertilizers, other stormwater discharges, etc. These sources of pollution severely contaminate the surface water, which further intensifies the use of groundwater.

8.2.2 Effects of Water Scarcity

The degradation of available water on the surface of the Earth (as depicted in Figure 8.3) has exacerbated the shortage of water. The disastrous effects due of water scarcity are:

1. Lack of access to good-quality drinking water
2. Starvation, poverty and negative impact on education

3. Unhygienic conditions of living — no proper sanitation
4. Spread of diseases
5. The forced migration of people to water-available areas
6. Destruction of habitat
7. Loss of biodiversity

Fighting the challenges associated with emerging pollutants in freshwater sources calls for urgent attention. These are initiated by controlling and prohibiting untreated waste discharges in water sources and minimizing water wastage,[2] to address existing and anticipated water shortages in many communities. Recycling or reusing water to protect water resources is an alternative option that can help communities significantly stretch their water necessities in an adequate, safe, and easily accessible manner. Thus, managing the impact of wastewater on the environment and attaining a sustainable level of water consumption requires the commitment and participation of many stakeholders.

8.3 Importance of Water Reuse

The water distribution on the Earth's surface is heterogeneous in nature, which has resulted in elevated water stress levels for centuries. The utilization of recycled water (domestic wastewater) has been documented since the beginning of the Bronze Age (around 3200–1100 BC), in Greek and Roman civilizations in cities of Egypt, Mesopotamia, and Crete. In the Minoan and Indus Valley (2600 BC) around 5,000 years before the Bronze Age civilizations, wastewater was utilized for irrigation and fertilization of agricultural lands. Throughout human history, the term "reuse" has referred to the use of raw or only partially treated wastewater for beneficial purposes, mostly for irrigation.

In recent decades, a fundamental change has taken place, as the advanced technologies of treatment plants have allowed treated water reuse to become an application for:

- Irrigation for agriculture
- Landscape irrigation such as parks and golf courses
- Process water for industrial processes such as power plants, refineries, mills, and factories
- Indoor uses such as toilet flushing
- Process water in construction sites for dust control or surface cleaning of roads
- Other construction processes such as concrete mixing
- Groundwater recharge or replenishing artificial lakes and inland aquifers
- Wetlands restoration
- Fire suppression
- Recreational water bodies

Water reuse, commonly known as water recycling or water reclamation, generally refers to the process of using treated wastewater for functional applications.[4,6] On a global scale, non-potable water reuse is overwhelmingly utilized in urban societies for beneficial purposes, including directly as potable water (dependent on the treatment technologies adopted in the water/wastewater treatment site). One of the key aspects of planning and designing a water reuse scheme is the quality of the influent wastewater and the purpose of the effluent wastewater. Different uses of water determine both the quality and quantity of the water available.

Alternative sources of water can substitute for existing water supplies and be used to enhance water security, sustainability, and resilience. However, public perceptions of reused water as a wasted or unclean resource, unsuitable for public or personal use, still persist.

8.4 History of Water Standards and Regulations

8.4.1 *Guidelines for Drinking Water Quality*

The origin of WHO Guidelines for Drinking Water Quality goes back to the 1950s.[1] They were intended for establishing national standards that will ensure the safety of drinking water, keeping the level of hazardous contamination in water to a minimum concentration. The standards or permissible limits were set using a risk-benefit approach by national authorities, taking into consideration the local environmental, economic, social, and cultural conditions of a region. Similarly, the Safe Drinking Water Act in 1974, with amendments in 1986 and 1996, defines the standards for drinking water quality and compels states, local authorities, and water suppliers/users to meet the required standards. The United States Environmental Protection Agency (EPA) has set maximum contaminant levels for over 90 different contaminants (including disease-causing germs and other organic-chemical contamination) viable to be present in public drinking water.

8.4.2 *Guidelines for Reclaimed Water Quality*

To protect public health, considerable efforts have been initiated to create regulations that would allow safe use of reclaimed water for various applications as highlighted in Section 8.3. The safety of water reuse is defined by the acceptable level of risks related to epidemiological occurrence.

International Reuse Guidelines was suggested by:[2–5]

- WHO 1989 Guidelines for the safe use of wastewater in agriculture and aquaculture — Guidelines for safe recreational water environments: Guidelines for drinking-water quality
- US EPA 1992 & 2004 for Reuse Guidelines: Safe Drinking Water Act.

An understanding of the nature/characteristics of wastewater is essential in the design and operation of collection, treatment, and disposal facilities and the engineering management of environmental quality. This requires knowledge of the actual amounts of physical, chemical, and biological constituents present in wastewater (Figure 8.4).

Untreated effluent water from sewage treatment plants, if discharged into surface water sources, would cause bacterial, viral, and parasitic diseases due to microbial contamination. WHO reported about 9.1% of the global burden of disease and 6.3% of all deaths (nearly 1 million people) are due to the most commonly occurring diseases *viz.* bacillary dysentery, infectious hepatitis, cholera, paratyphoid, typhoid, giardiasis, salmonellosis, and cryptosporidiosis, transmitted through drinking of

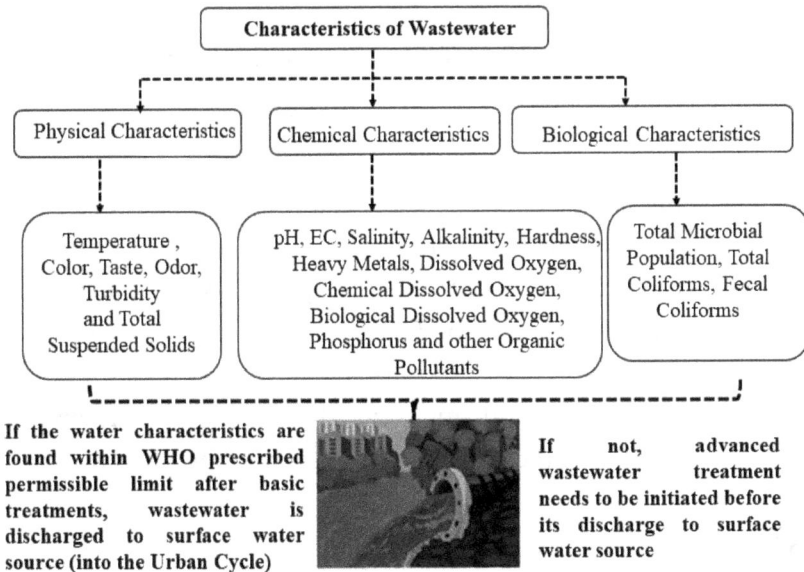

Figure 8.4   Basic attributes of wastewater.

unsafe water. Worldwide, 11% of child deaths are attributed to diarrheal disease, 88% of which are caused by unsafe water or improper sanitation and poor hygiene. It has been estimated that annually around 37.7 million are affected by, and 73 million working days lost due to, waterborne diseases. Since access to safe drinking water is important as a health and development issue, investments in water supply and sanitation can yield a net economic benefit, owing to reductions in adverse health effects and health care costs. Thus, improving the quality of water sources, adequate access to safe water, safe storage, adequate sanitation facilities, treating household water, and encouraging good hygiene practices can ensure good public health. In recent years, the need and benefits of using both freshwater and reclaimed water in regions with extreme water shortages have been well recognized and documented worldwide.

8.5 Advances in Treatment Technologies of Water

Advances in treatment technologies have made it possible to recycle water to a quality that is fit for all public and personal uses; however, public perceptions still exist that reused water is a spoiled or unclean resource, unsuitable for public or personal use. Therefore, wastewater managers should employ holistic and comprehensive risk assessment and management techniques/approaches in dealing with water pollution issues to ensure the safety and sustainability of all aquatic systems, and ensure that the effluent that is eventually released into the environment does not degrade the quality of the recipient water bodies.

8.5.1 *Sequence of Wastewater Treatment*

The technologies and their sequence of use in the treatment of wastewater include (Figure 8.5):

(a) Preliminary treatment of wastewater
(b) Primary treatment of wastewater

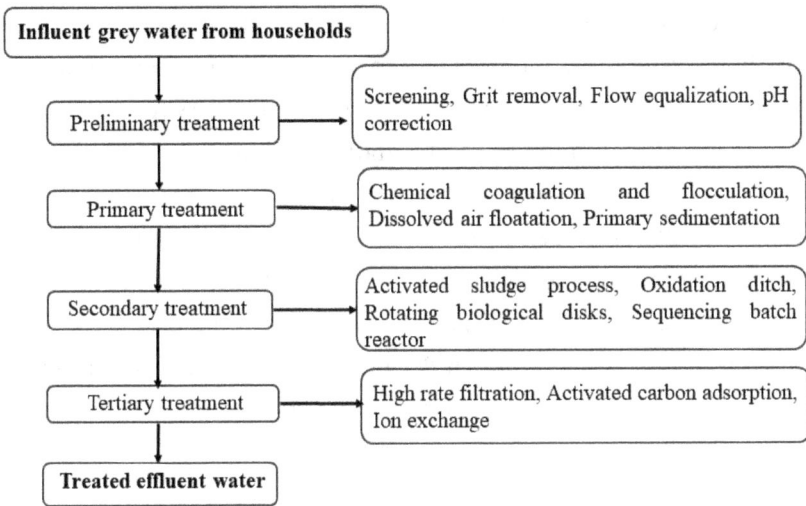

Figure 8.5 Sequence of wastewater treatment.

(c) Secondary treatment of wastewater

(d) Tertiary treatment of wastewater

Preliminary and Primary treatment of wastewater generally include those processes that remove debris (such as rags, grit, sticks, and other foreign objects) and coarse biodegradable material from the waste stream and/or stabilize the wastewater by equalization process. Grit removal, compared to other unit treatment processes, is quite economical and employed to achieve the following results:

– Restricts excessive abrasive wear of equipment such as pumps and sludge scrapers.
– Prevents accumulation of grits in subsequent operating processes in channels, pipes, and basins.
– Reduces the capacity in sludge-handling facilities.

In wastewater treatment plants, common grit sizes range from 50 to 100 mesh, corresponding to 0.3 mm to 0.15 mm, with a specific gravity of 2.65 and a settling velocity of 20 mm/s. In domestic wastewater treatment, both preliminary and primary processes will remove approximately 25% of the organic load and non-organic solids. Primary treatment technologies are often limited to sedimentation with and without chemical addition. Plain sedimentation processes alone assist in removal of 30–40% of the biochemical oxygen demand (BOD),[a] as well as 40–70% of oil and 65% of grease from suspended solids, in addition to the removal of organic nitrogen, organic phosphates, and heavy metals in the process.

Secondary treatment of wastewater generally includes the removal of organic pollutants via biological processes. Biological processes can be broadly classified as:

i) Aerobic — microbes that require oxygen to grow
ii) Anaerobic — microbes that grow in the absence of oxygen but use other compounds such as sulfate, phosphate, or other organics present in the wastewater other than oxygen
iii) Facultative — microbes that can grow in the presence or absence of oxygen.

In an aerobic treatment plant, influent wastewater of 100 kg chemical oxygen demand (COD)[b] load requires the energy of 100 kWh for aeration and produces 30–60 kg of sludge, with the outgoing effluent having a COD load of 2–10 kg. However, in an anaerobic treatment plant, for the same 100 kg load COD, the sludge production is only 5 kg, and energy recovery is 40–45 m^3 (biogas) and 382 kWh of electricity. BOD_5 removal is between 85–95%.

Tertiary treatment of wastewater generally includes advanced treatment processes for further removal of organics using materials like activated carbon, ozone, membranes, etc. These techniques assist in further disinfection of wastewater, decolorization, odor removal, and organic

---

[a] The biochemical oxygen demand is the amount of oxygen required to completely oxidize organic compounds to carbon dioxide and water by microorganisms.
[b] The chemical oxygen demand refers to the amount of oxygen required to degrade all oxidizable compounds.

degradation. Turbidity reduction up to 99%, color up to 31–76%, surfactants up to 32–94%, and COD removal of 89–91% can be achieved. Further, advanced treatment processes are capable of also addressing contemporary water quality issues related to potable reuse, such as emerging pathogens or trace organic chemicals.

8.6 Significance of Circular Economy Concept in Wastewater Reuse

Since the availability of clean and safe water for potable and non-potable uses is declining on Earth's surface, many developing countries and emerging economies are focused on improving their waste management systems via the concept of circular economy. The transition to a circular economy in wastewater management is gripping most nations as it provides opportunities in terms of economic outputs and jobs, and widespread application of water reuse.

Under the directives of the Urban Wastewater Treatment Directive (91-271-EEC), Europe has already contributed to obtaining treated wastewater of quite high quality for potable uses. It has been estimated that global water reuse capacity has increased from 33.7 GL/day to 54.5 GL/day from 2010 to 2015, with the largest developments in China, the United States, the Middle East, North Africa, Europe, and South Asia. It is expected that with the move towards *circular economy*, this reuse will further accelerate. Table 8.2 lists the indirect reuses of wastewater along with the necessary water quality parameter to be monitored.

8.6.1 Challenges in Wastewater Reclamation and Water Circularity

During the past decades, treatment technologies for wastewater reclamation have made substantial progress; however, the overall reclamation ratio of treated wastewater is still low. There are a few possible reasons:

Table 8.2   Recommended water quality criteria for wastewater reuse.

Water Quality Parameter Wastewater Reuse	BOD₅, mg/L	Total Suspended Solids, mg/L	Turbidity, NTU	Feacal Coliform, /100 ml	Total Nitrogen, mg/L	Total Phosphorus, mg/L	pH	Drinking Water Standards
Crop irrigation	30	50	20	1,000	30	6–8	6–9	No
Reservoir recharge	≤3	≤5	5	≤3	≤10	0.6	6–9	Yes
Groundwater recharge	≤3	≤5	10	≤500	≤10	≤1	6–9	Yes
Domestic non-potable	≤3	≤5	≤2	≤500	≤10	≤1	6–9	No

1. Lack of public perception and acceptability.
2. Lack of outreach and promotion for the aquifer recovery application.
3. Lack of resources allocated to wastewater reclamation in total urban planning.
4. Incomplete regulations and supporting policies for wastewater reclamation and replenishment.
5. Clogging of underground sources due to replenishment of aquifers with partially treated water.

As freshwater supplies are limited, technological interventions in the wastewater treatment process have helped in economic development in many regions. A transition to a *circular economy*, where recycling and reuse are central themes, can further help handle the growing imbalance between water demand and supply.[6]

Although water reuse faces numerous barriers, ranging from public perception to high prices associated with advanced technological interventions, a combination of approaches such as the conservation of water, recycling, and treatment of wastewater from non-traditional resources to "create" new water is the light at the end of the tunnel. It can bring water circularity while reducing the cost of water (Figure 8.6).

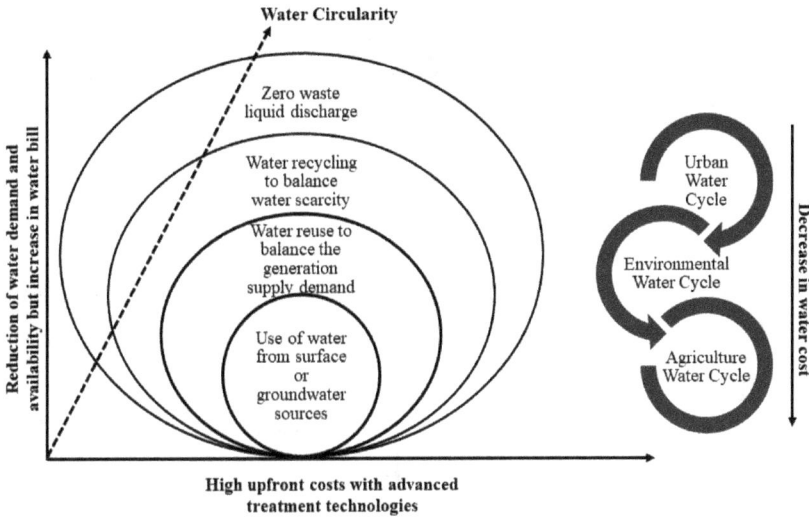

Figure 8.6   Water circularity and associated cost strategies.

8.6.2 *A Successful Case Study in Water Sustainability: Singapore*

Singapore — once considered one of the most water-stressed countries in the world due to its limited land and natural resources — has successfully demonstrated a sustainable water supply system. This achievement is primarily attributed to its Four National Taps strategy, which includes:

1. Local catchment water,
2. Imported water,
3. NEWater (high-grade reclaimed water), and
4. Desalinated water.

Singapore has adopted robust and innovative technologies based on four key pillars, as illustrated in Figure 8.7, to harness, treat, and reuse every drop of water. In this context, public–private partnerships play a crucial role in the conservation and sustainable management of water resources.[1] By implementing a circular water management system, Singapore has been able to minimize the cost of water. Water pricing is

PILLAR 1:
WATER AND SUSTAINABLE MANAGEMENT

Addresses the importance of resource sustainability as a major ingredient for PUB's mission and PUB's responsibility in managing its environmental impacts

· Sustainable Water System
· Resource Efficiency and Circularity
· Climate Change Adaptation

PILLAR 2:
CAPABLE AND ENGAGED WORKFORCE

Addresses the importance of our people as the main driving force and cornerstone of PUB's operations

· Health and Safety
· Competent Workforce
· Inclusive and Fair Workplace

PILLAR 3:
STRONG PARTNERSHIPS

Addresses the importance of the shared responsibility of all stakeholders in the community to conserve and protect our water resources and environment

· Customer-centric Water Service
· Partnership and Engagement

PILLAR 4:
BUSINESS EXCELLENCE

Addresses the importance of business excellence to ensure a PUB that is responsible and trusted by all our stakeholders in delivering our mission

· Trust and Transparency
· Innovation
· Digitalisation and Cybersecurity
· Financial Sustainability

Figure 8.7   Water sustainability framework in Singapore. Adapted with permission.[1]

based on household consumption, and to ensure affordability, the government provides rebates on public water bills. Furthermore, institutions such as the Singapore Water Academy and Public Utilities Board are actively working toward achieving the goals outlined in Singapore's Green Plan 2030, which emphasizes:

- Water quality,
- Sustainable urban stormwater management,
- Water reuse, and
- Efficient supply network management.

In addition, the Public Utilities Board collaborates with UNICEF to support the global Water, Sanitation, and Hygiene (WASH) agenda, extending its commitment to sustainable water practices beyond national borders.

References

1. Arora. P, (2018) Physical, Chemical and Biological Characteristics of Water. Environ Sci 3: 4–16.

2. EPA (2001) Source Water Protection Practices Bulletin Managing Septic Systems to Prevent Contamination of Drinking Water, EPA 21: 1–3.

3. WHO (2017) Guidelines for drinking-water quality. WHO: Geneva.

4. World Water Development Report (WWDR). Water for People, Water for Life; The United Nations Educational, Scientific and Cultural Organization: Paris, France, 2003.

5. American Water Works Association. 1994. Dual Water Systems. McGraw-Hill. New York.

6. https://www.pub.gov.sg/sustainability

CHAPTER NINE
EMERGING NANOMATERIALS
FOR SUSTAINABILITY

9.1 Introduction

Nanomaterials are materials that possess at least one dimension smaller than 100 nm. They may occur naturally, form as by-products of combustion reactions, or be intentionally engineered to perform specific functions.

The European Commission has defined nanomaterials as *"the particle size of at least half of the particles in the number size distribution must measure 100 nm or below to be regarded as nanomaterial"*.

Engineered nanomaterials, characterized by their unique physical and chemical properties, have garnered significant attention due to their high tunability across optical, electronic, magnetic, tribological, and chemical domains. These exceptional properties offer transformative potential in various sustainability-focused technologies, including microelectronics, spintronics, medical applications, and other advanced devices.[1] The versatility and adaptability of nanomaterials make them

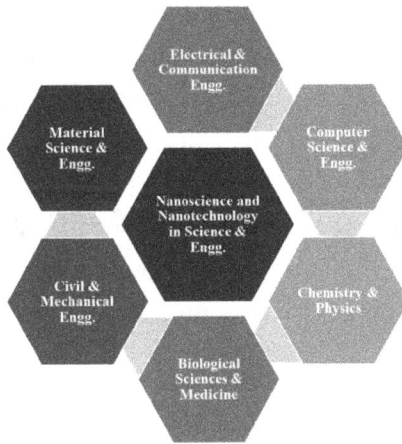

Figure 9.1   Application of nanomaterials supporting sustainability in different fields.

valuable in building sustainability across multiple sectors (Figure 9.1), as outlined below:

(a) Industrial Applications: Nanomaterials are in high demand in the industrial sector due to their durability, porosity, and enhanced mechanical properties, all of which contribute to material sustainability.

(b) Commercial Use: For decades, nanomaterials have been incorporated into commercial products such as stain-resistant and wrinkle-free textiles, cosmetics, sunscreens, paints, coatings, and varnishes, supporting the production of sustainable consumer goods.

(c) Consumer Products: Nanomaterials are widely used in items such as sports equipment, bicycles, automobiles, and protective coatings, offering lightweight strength and durability.

(d) Energy Sector: Nanomaterials are revolutionizing energy storage, conversion, and efficiency. For example, nanostructured stained electrodes in solar cells increase the optical absorption coefficient, leading to improved energy conversion efficiency, thereby supporting sustainable energy solutions.

(e) Electronics and Computing: In electronics and computing, nano-materials help improve the precision and performance of circuits at the atomic level, supporting the development of compact, energy-efficient, and high-performance devices.

(f) Medical Applications: In the medical field, nanomaterials are valu-able due to their large surface-to-volume ratio, allowing for efficient bonding with cells and active compounds. This facilitates targeted drug delivery and improved treatment of various diseases.

9.2 Types and Classification of Nanomaterials

Nanomaterials are classified into three main categories as shown in Figure 9.2.

(a) Organic-based nanomaterials: These nanomaterials consist of par-ticles with two or more dimensions, with a size in the range of

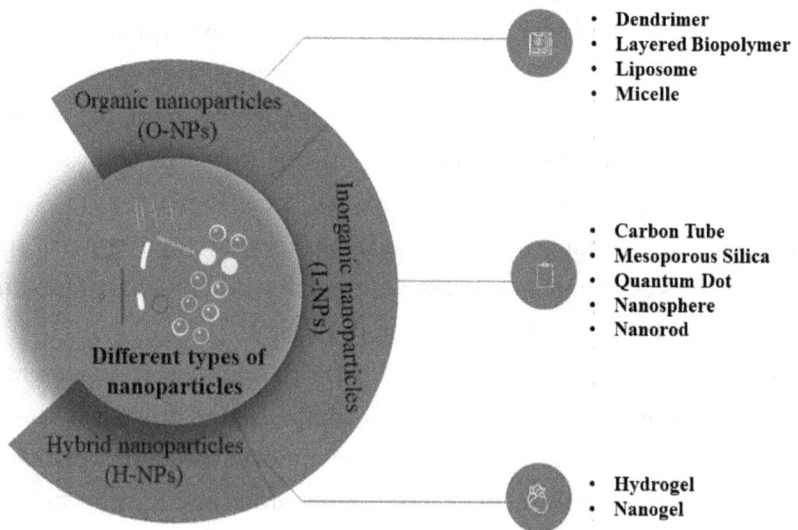

Figure 9.2   Different types of nanomaterials.

1–100 nm. The major groups include liposomes, micelles, biopolymer/ protein-based layers, and dendrimers. Dendrimers, in particular, are widely used for targeted drug delivery and tissue-repair scaffolds due to their unique branched synthetic polymer structure (typically <15 nm). They have a layered architecture composed of a central core, an internal region, and numerous terminal groups.

(b) Inorganic-based nanomaterials: This category primarily includes nanoparticles such as carbon nanotubes, which belong to the fullerene family. These are cylindrical or tubular structures (<100 nm) made from coaxial graphite sheets. Due to their metallic or semiconductor properties, they are effective as biosensors, drug carriers, and tissue repair scaffolds. Other examples include quantum dots (2–10 nm), often used in imaging and bioassay labeling, and nanospheres and nanorods with inorganic cores that exhibit magnetic, electric, fluorescent, and optical properties.

(c) Hybrid-based nanomaterials: Hybrid nanomaterials are unique chemical conjugates combining organic and/or inorganic components. Examples include perovskites, sol-gel silica modified with organic molecules, active organic molecules doped into conductive polymers, organically modified mesoporous materials, mixed organic-inorganic polymers, silica plus organic polymer, polymer-magnetic nanoparticle composites, polymer-coated inorganic nanoparticles, and silica-embedded bioactive species.

Based on the dimensions, nanomaterials can be placed into four different classes as shown in Table 9.1. They can exist in single, fused, aggregated, or agglomerated forms with spherical, tubular, and irregular shapes.

9.3 Synthesis of Nanomaterials

Generally, three methods are used to synthesize nanomaterials: (1) Physical or mechanical route (top-down approach), (2) chemical

Table 9.1 Classification of nanomaterials based on dimension.

Dimension	Characteristics	Nanomaterials
Zero	Nanomaterials with all their three dimensions in the nanoscale range. Example: quantum dots, nanoparticles.	
One	Nanomaterials with one dimension outside the nanoscale. Example: nanotubes, nanowires, nanofibers, nanorods, etc.	
Two	Nanomaterials with two dimensions outside the nanoscale. Example: nanosheets, nanofilms, etc.	
Three	Nanomaterials confined to the nanoscale in any dimension. Example: bulk powders, dispersions of nanoparticles, etc.	Dispersion Aggregation

route (bottom-up approach), and (3) biological route, as depicted in Figure 9.3.

9.3.1 Physical/Mechanical Route (Top-down Approach)

This method involves breaking down bulk materials into smaller, nano-sized particles. Techniques include mechanical milling, etching, ablation, sputtering, and electro-explosion, where mechanical tools are used to cut, mill, and shape materials to the desired size and form. Mechanical milling is a cost-effective method for producing nanoscale materials from bulk sources. It is especially useful for creating blends of different phases and for manufacturing nanocomposites.

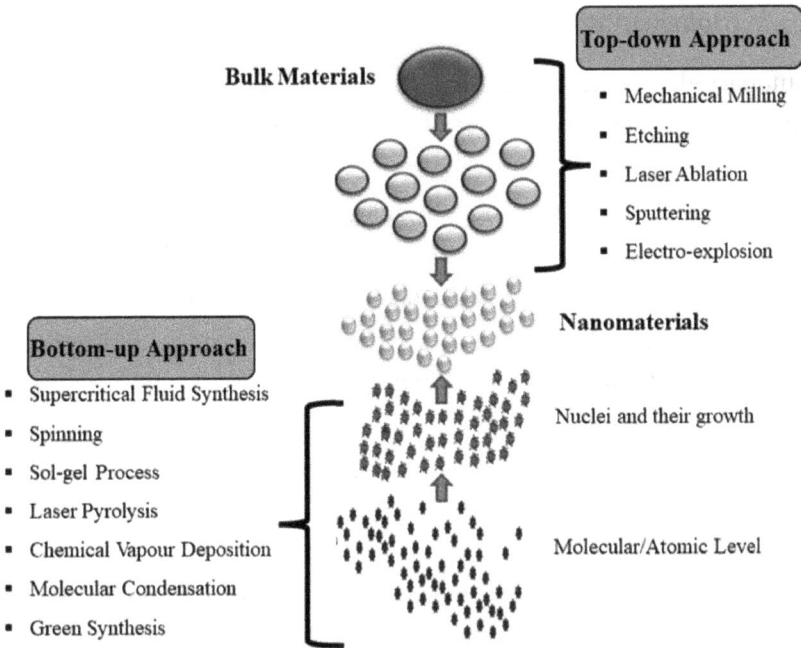

Figure 9.3 Synthesis of nanomaterials using top-down and bottom-up approaches. Reproduced with permission.

9.3.2 Chemical Route (Bottom-up Approach)

The synthesis method plays a crucial role in determining the characteristics of the nanoparticles. Several wet chemical methods have been widely used over the past few decades, including sol-gel, polyol, electrochemical, citrate precursor, sonochemical, solvothermal, co-precipitation, and hydrothermal methods. However, factors such as reagent composition, reaction conditions, energy consumption, and production costs can significantly influence the properties of the resulting nanoparticles.

9.3.3 *Biological Route*

Nanoparticles synthesized by biological routes are called bio-nanoparticles. These are assemblies of atoms or molecules formed within biological systems, with at least one dimension ranging from 1 to 100 nm. Bio-nanoparticles such as gold, silver, cadmium, copper, platinum, titanium, palladium, and zinc are produced by reducing, capping, or stabilizing microorganisms like bacteria, actinomycetes, fungi, yeasts, viruses, and algae. These bio-nanoparticles have various applications in industrial and biomedical fields. Biodegradable materials such as organic acids, proteins, vitamins, and secondary metabolites — including flavonoids, alkaloids, polysaccharides, terpenoids, heterocyclic compounds, and polyphenol-rich agricultural residues — act as reducing and capping agents without toxic effects.

Bio-nanoparticles are categorized into (a) intracellular and (b) extracellular structures. Whether nanoparticles are synthesized intracellularly or extracellularly mainly depends on specific changes in ionic conditions and temperature during synthesis. Nanoparticles formed inside cells tend to be smaller, making them suitable for more specialized applications. In contrast, extracellularly synthesized nanoparticles can be readily utilized in areas such as optoelectronics, bioimaging, electronics, and sensor integration.

Plant extracts (from roots, fruits, stems, seeds, leaves, or polyphenol-rich agricultural residues) are also used as bio-based sustainable materials for nanoparticle synthesis. Since fungi can secrete large amounts of proteins, they are generally more effective than bacteria for producing larger quantities of bio-nanoparticles.

9.4 Application of Nanomaterials in the Biomedical Field

Nanomaterials have gained significant attention and research interest in a wide range of biomedical applications (Figure 9.4). Their beneficial properties — such as high chemical stability, non-toxicity,

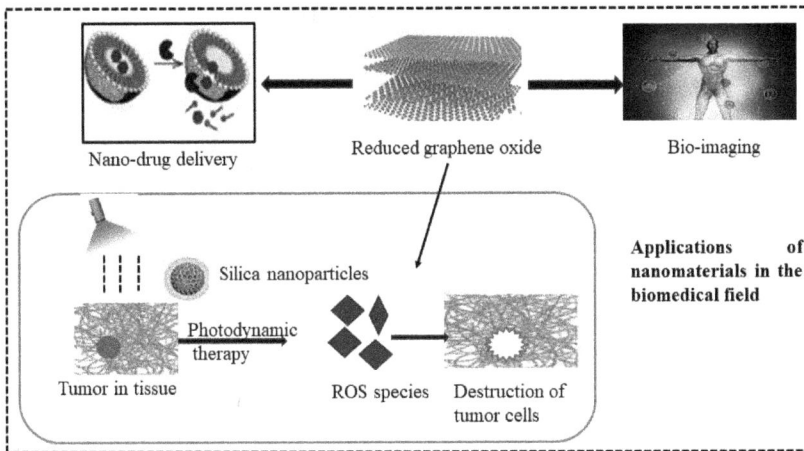

Figure 9.4 Nanomaterial applications in biomedical fields.

biocompatibility, cost-effectiveness, and environmental friendliness — have led to their widespread use in imaging, cancer therapy, diagnosis, targeted drug delivery, and hyperthermia treatments. Commonly used nanomaterials in healthcare include carbon nanotubes, nanofibers, and reduced graphene oxide. These nanomaterials are synthesized by physical, chemical, and biological methods.

Graphene-based nanomaterials are particularly sustainable and widely used in modern diagnostic techniques. It has been recently reported that, in the presence of 660 nm laser excitation, nanographene oxide materials help in the detection of B-lymphocyte cells.[2] Similarly, graphene quantum dots (3–5 nm) assist in the recognition of HeLa (immortal human cell lines) cells in the cytoplasm, enabling early tumor detection in organs.[3] Furthermore, mesoporous silica nanoparticles are being used in photodynamic therapy to destroy cancerous cells and harmful pathogens. It is a non-invasive treatment strategy that is based on the principle of photosensitization. In this therapy, nanomaterials act as photosensitizers that absorb photons at the wavelength of 470 nm, producing highly reactive oxygen species (ROS) inside the cell. These ROS radicals are capable of destroying tumorigenic and malignant cells.

Nanomaterials have also shown promise as targeted drug delivery agents. Notable nanomaterials with excellent drug-carrying capacity include platinum, selenium, and silver. For instance, silver nanoparticles synthesized from *Cryptococcus* species exhibited anti-cancer properties against breast cancer cells.[4] Selenium nanorods, synthesized from *Streptomyces* species, initiated the destruction of hepatocellular carcinoma cells. Research suggests that the selenium nanorods bind to the DNA and copper (II) ions to form a ternary complex. The nanorods from the ternary complex help in the electron transfer which results in the reduction of copper (II) ions to copper (I). This eventually generates ROS species which mediate the selective killing of the cancerous cells. Other studies report the use of zinc oxide nanoparticles synthesized from *Rhodococcus* species to successfully deliver anthraquinone for carcinoma treatment.[5] Similarly, biosynthesized gadolinium oxide nanoparticles have been used to deliver doxorubicin for cancer therapy.[6]

Despite these benefits, challenges remain with nanomaterial-based drug delivery. It has been found that sometimes these nanoparticles can alter the cell membrane and cause damage to untargeted cells. For example, copper oxide nanoparticles synthesized from *Sargassum* species have reportedly caused an increase in the levels of serum creatinine in the body.[7] Similarly, there are reports of accumulation of zinc and palladium nanoparticles in the liver, lungs, and kidney.[8] Therefore, in order to overcome these challenges, significant research needs to be done to develop greener, safer nanomaterials that minimize harmful residues and side effects within the body.

9.5 Applications of Nanomaterials in Wastewater Treatment

The need for clean and potable water is increasing rapidly due to urbanization and industrialization. Conventionally, wastewater from industries, agricultural, household, and commercial sectors was treated using various physical, chemical, and biological methods. However, in recent years, nanotechnology has gained attention as a cost-effective and environmentally friendly alternative for wastewater treatment.

Nanomaterials, which are smaller than 100 nm in diameter, possess a large surface-to-mass ratio and high adsorption capacity, making them highly effective for treating wastewater. Besides conventional metallic and non-metallic nanomaterials, green nanomaterials — synthesized from plant extracts, microorganisms, and various biowaste products (such as vegetable and fruit peels, and eggshells) — are among the safest, most affordable, and ecologically acceptable options for environmental remediation. These greener nanomaterials are processed into various forms such as nanotubes, films, nanowires, and colloids, which are then used to remove pollutants (Figure 9.5).

Titanium dioxide (TiO_2) nanomaterials are sustainable and widely used for the removal of heavy metals and ions such as arsenic, lead, and chromium from wastewater. In the presence of sunlight, TiO_2 nanoparticles convert chromium (VI) to the less harmful chromium (III), thereby

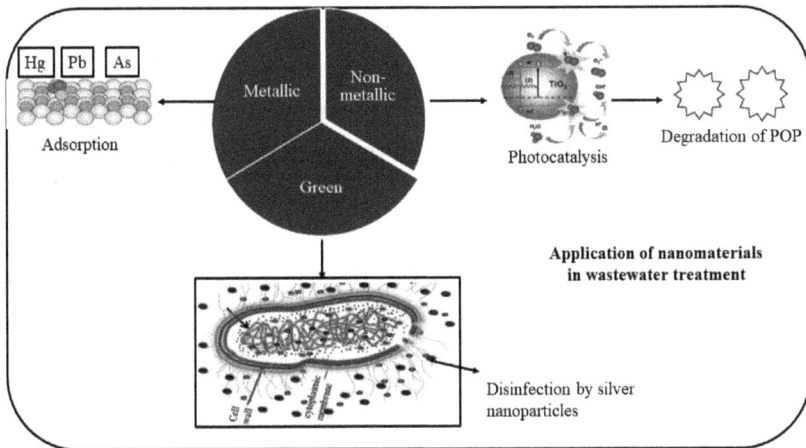

Figure 9.5   Nanomaterial applications in wastewater treatment.

reducing the toxicity of the water sample. Similarly, divalent heavy metal ions like cobalt, nickel, and copper are removed by iron-oxide nanocomposites acting as adsorbents, with effective removal occurring between pH 3 and 7.[9] Adsorption on nanomaterials is considered a fast, reliable, and cost-effective method for pollutant removal. Other nanomaterials like carbon nanotubes, manganese oxide, and zinc oxide are extensively used for adsorbing pollutants, which also helps improve water permeability by reducing wastewater salinity.

Furthermore, persistent organic pollutants such as polychlorinated biphenyls, chlorinated alkanes, and industrial chemicals from wastewater are effectively degraded/removed in the presence of nanocatalysts. It has been reported that when graphene is doped in TiO_2, it enhances pollutant breakdown by generating a greater number of hydroxyl radicals.[10] Zerovalent iron nanoparticles are very efficient in degrading chlorinated organic molecules. Nanomaterials such as silicon dioxide, superparamagnetic iron oxide nanoparticles, and magnetic carbon nanotubes enhance degradation rates through interactions between bacteria and nanoparticles.

Nanomaterials are also used as nanomembranes for the removal of micropollutants. Electrospun nanofibrous membranes have proved to be a promising technology for the removal of nanoplastics, which are 50 nm[11] in size, with 90% efficiency. In this method, nanofibers are placed between two membranes which increases the efficiency of separation. The positively charged nanofibers attract negatively charged nanoparticles onto their surface, trapping them inside the membrane.

Additionally, green nanomaterials such as silver nanoparticles synthesized from *Cucumis prophetarum* leaves serve as antibacterial and antifungal agents. These nanoparticles form silver radicals (Ag^+) that alter the cell membrane after adhering to the surface of bacterial cell walls, thereby deactivating the bacterial enzymes. This ultimately results in the death of bacterial pathogens present in the wastewater. Thus, silver nanoparticles play a significant role in wastewater treatment.

9.6 Application of Nanomaterials in Energy Storage Devices

Another important application of nanomaterials is in energy storage devices such as batteries. Lithium-ion (Li-ion) batteries are widely used in various electrical and electronic appliances, including electric vehicles, because they are considered safe, offer high energy density, and have a long life cycle. The basic components of Li-ion batteries include particles sized between 10 to 50 μm deposited on aluminum and copper current collectors (cathode and anode) along with electrolytes as conductivity enhancers and binders/separators. The separator is comprised of a porous membrane that is soaked in an electrolyte solution, which allows Li ions to flow between the electrodes and thus avoid short circuit. The anode typically contains carbon-based materials, such as graphite and graphene. Battery performance is evaluated based on energy, power, life cycle, cost, and thermal stability.[12]

Recently, different varieties of cathodes and anodes have been developed. Cathodes include mixed metal oxides like $LiMn_{1.5}Ni_{0.5}O_4$ (spinel), $LiNi_{0.33}Mn_{0.33}Co_{0.34}O_2$ (layered), and $LiNiCoAlO_2$ (layered), as well as metal phosphates like $LiFePO_4$, $LiCoPO_4$, and $LiMnPO_4$ (olivine). Anodes include lithium titanates ($Li_4Ti_5O_{12}$) and tin oxide (SnO_2), elemental silicon (Si), tin (Sn) and many carbon-based nanomaterials.

The significance of the electrolyte solution is to provide high ionic conductivity between the electrodes. However, it is highly reactive, flammable, toxic, and can leak out of the battery. Another issue is the formation of a passivating layer called the solid-electrolyte interphase, which can cause unwanted reactions with the anode. Thus, to solve these issues, conventional separators/electrolytes are replaced with solid electrolytes — both organic (polymer-based) and inorganic — to obtain solid-state batteries, as represented in Figure 9.6. Solid-state batteries help stabilize the interface between the electrolyte and both metallic lithium and the cathode. Fluorinated electrolytes are used in some batteries to form a stable interphase between the solid electrolyte anode and the cathode, enabling stable, high-energy density Li-ion batteries.

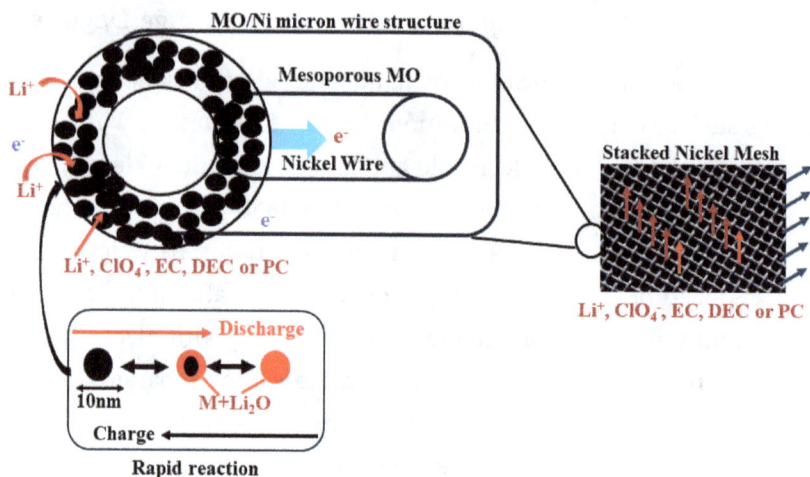

Figure 9.6 A schematic representation of a solid-state lithium storage device. Adapted with permission.[13]

References

1. Wiederrecht, G. P., Bachelot, R., Xiong, H., Termentzidis, K. et al., (2023) Nanomaterials and sustainability, ACS Energy Letters, 8, 3443–3449.

2. Semenov, K. N., Shemchuk, O. S., Ageev, S. V., Andoskin, P. A. et al. (2024) Development of graphene-based materials with the targeted action for cancer theranostics, Biochemistry, 89, 1362–1391.

3. Lannazzo, D., Espro, C., Celesti, C., Ferlazzo, A., et al. (2021) Smart biosensors for cancer diagnosis based on graphene quantum dots, Cancers (Basel), 13(13), 3194. doi: 10.3390/cancers13133194.

4. Ortega, F. G., Fernández-Baldo, M. A., Fernández, J. G., Serrano, M. J., et al. (2015) Study of antitumor activity in breast cell lines using silver nanoparticles produced by yeast, International Journal of Nanomedicine, 10, 2021–2031. doi: 10.2147/IJN.S75835.

5. Anjum, S., Hashim, M., Malik, S. A., Khan, M., et al. (2021) Recent advances in zinc oxide nanoparticles (ZnO NPs) for cancer diagnosis, target drug delivery, and treatment, Cancers (Basel), 13(18), 4570. doi: 10.3390/cancers13184570.

6. Babayevska, N., Kustrzyńska, K., Przysiecka, L., Jarek, M., et al. (2025) Doxorubicin and ZnO-loaded gadolinium oxide hollow spheres for targeted cancer therapy and bioimaging, Spectrochimica Acta Part A: Molecular and Biomolecular Spectroscopy, 333, 125903. doi: 10.1016/j.saa.2025.125903.

7. Ramírez, S. J. F., Morales, B. E., Velueta, D. A. P., Grajeda, J. M. T. S., et al. (2024) Green synthesis of copper nanoparticles using *Sargassum spp.* for electrochemical reduction of CO2, ChemistryOpen, 13(5), e202300190. doi: 10.1002/open.202300190.

8. Aarzoo, Siddiqui M. A., Hasan, M., Nidhi, et al. (2024) Palladium nanoparticles and lung health: assessing morphology-dependent subacute toxicity in rats and toxicity modulation by naringin – paving the way for cleaner vehicular emissions, ACS Omega, 9(30), 32745–32759. doi: 10.1021/acsomega.4c02269.

9. Sibiya, N. P., Mahlangu, T. P., Tetteh, E. K. and Rathilal, S. (2024) Review on advancing heavy metals removal: The use of iron oxide nanoparticles and microalgae-based adsorbents, Cleaner Chemical Engineering, 11, 100137. doi: 10.1016/j.clce.2024.100137.

10. Cruz, M., Gomez, C., Duran-Valle, C. J., Pastrana-Martínez, L. M., et al. (2017) Bare TiO2 and graphene oxide TiO2 photocatalysts on the degradation of selected pesticides and influence of the water matrix, Applied Surface Science, 416, 1013-1021. doi: 10.1016/j.apsusc.2015.09.268.

11. Pervez, M. N., Mishu, M. R. and Naddeo, V. (2022) Electrospun nanofiber membranes for the control of micro/nanoplastics in the environment, Water Emerging Contaminants & Nanoplastics, 1, 10. doi: 10.20517/wecn.2022.05.

12. Babbitt, C. W. and Moore, E. A. (2018) Sustainable nanomaterials by design: a new material selection tool offers a proactive perspective on nanotechnology research and development, Nature Nanotechnology, 13, 621–629.

13. Jiang, C., Hosono, E. and Zhou, H. (2006) Nanomaterials for lithium ion batteries, Nanotoday, 1, 28–33.

CHAPTER TEN
LIFE CYCLE THINKING FOR MATERIALS SUSTAINABILITY

10.1 Introduction

Life Cycle Thinking (LCT) considers the potential environmental, economic, and social impacts at every stage of a product or service's life cycle, with the goal of minimizing negative outcomes. Through this approach, sustainability-related issues — such as greenhouse gas emissions, material availability, job creation, and product or service longevity — can be identified and addressed. LCT is generally built on three key forms of life cycle assessment (LCA):

- Environmental LCA
- Social LCA
- Life cycle costing

LCT is commonly applied in product design, manufacturing processes, and strategic planning, helping to assess each phase of the life cycle — such as raw material extraction, material processing, manufacturing, transportation, reuse, recycling, and disposal — as illustrated in Figure 10.1. LCT also sheds light on consumer behavior and how it

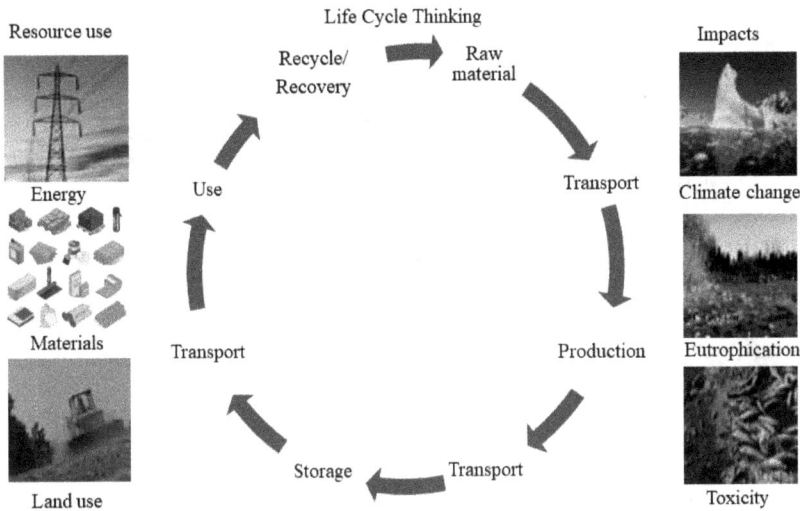

Figure 10.1 Life cycle thinking approach for any product or service.

affects sustainability. Under a linear economy model, a product follows a simple path: buy → use → dispose. At the end of its life, the product typically ends up either in a landfill or, less frequently, is recycled.[1] Products such as plastic bags, water bottles, batteries, mobile phones, and computers often have short life spans and contribute significantly to waste accumulation. This linear model causes hidden environmental, social, and economic impacts, which underscores the need for a broader perspective — like LCT — to better predict, evaluate, and manage these impacts.

For instance, the linear economy model for mobile phones, as shown in Figure 10.2, illustrates environmental impacts from mining through to landfilling. It requires an LCT approach to minimize the impacts. Generally, the lifespan of a mobile phone ranges from 15 to 18 months, though this can vary depending on geographical location and user behavior. At the end of its life, mobile phones contribute significantly to electronic waste. By applying the LCT approach, mobile phones can be dismantled and separated into individual components — such as plastic

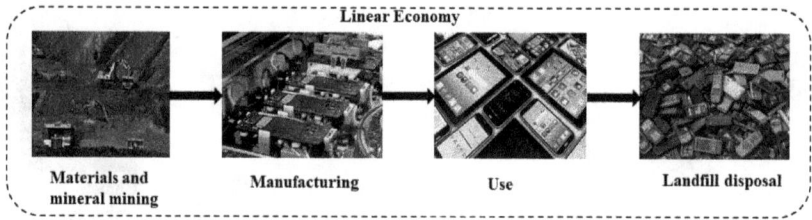

Figure 10.2   Linear economy approach in the life cycle of mobile phones.

casings, screens, and batteries — maximizing the recovery of valuable materials and reducing environmental harm.

10.2 Life Cycle Assessment

To build sustainability in materials, it is essential to evaluate both the environmental and economic impacts of products throughout their life cycle. LCA is a widely accepted method used to interpret these impacts and address environmental concerns. It is applied across various sectors — including corporate industries, businesses, and government agencies — for product development and policymaking.

"The life cycle assessment aims to evaluate and quantify the environmental effect of a product/process/service via a given systemic protocol originating from extraction, manufacturing, process creation, usage, and disposal, through assessing the entire life cycle." –ISO 14040

LCA studies began in the 1960s, originally focused on pollution prevention. At that time, no standardized procedures or guidelines were available, and different approaches were used depending on the specific issue being addressed. In the early 1970s, several companies began analyzing the environmental impact of their products using LCA methods. However, due to the limited availability of primary data, early assessments were largely based on secondary data sources.

10.3 Life Cycle Assessment Stages

As shown in Figure 10.3, LCA consists of five key stages:

1. Raw material extraction
2. Manufacturing and packaging
3. Transportation
4. Use and recycling
5. Waste management or end-of-life

To conduct an LCA study effectively, four standardized phases are defined:

1. Goal and scope definition
2. Inventory analysis

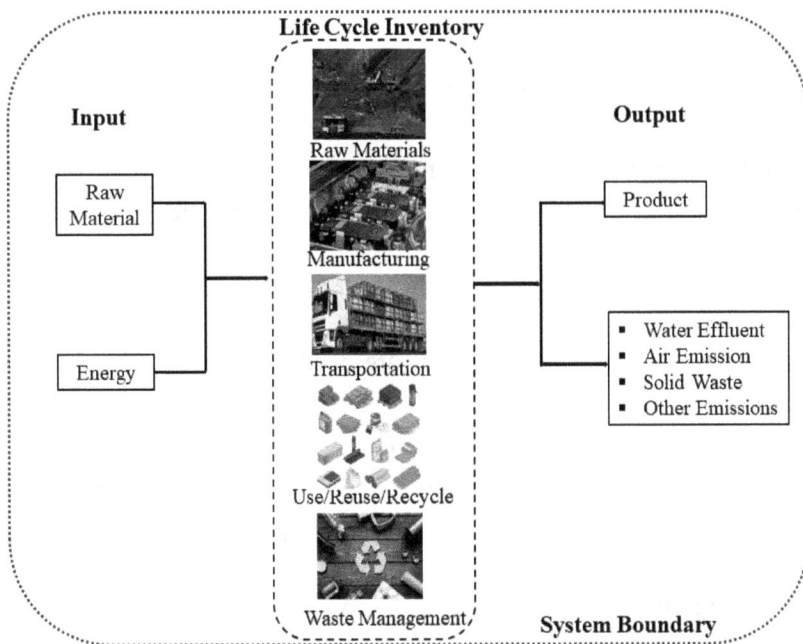

Figure 10.3   An overview of an LCA of a product.

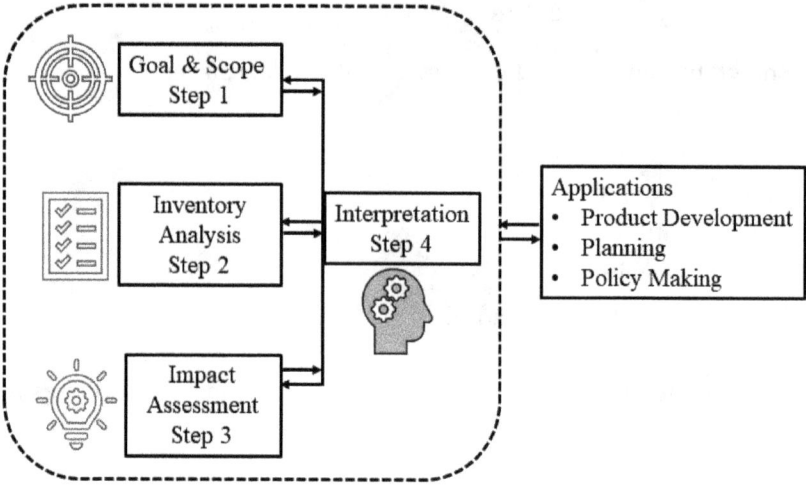

Figure 10.4   LCA framework for materials.

3. Impact assessment
4. Interpretation

The relationship between these four phases is interconnected, as shown in Figure 10.4, and is established under ISO 14040 and ISO 14044 standards. These phases are essential for systematically analyzing the environmental attributes of a product throughout its life cycle.

10.3.1 *Goal and Scope*

Qualitative and quantitative details are needed for LCA, therefore the first step involves establishing the goal of the study. Following this, the scope must be outlined by defining the functional unit, product system, and system boundaries for input and output flows. During the scoping phase, several key questions are typically addressed:

- What should or should not be assessed?
- What system is being evaluated?
- What is the functional unit?

- What are the boundary limits of the analysis?
- What type of life cycle is being considered (e.g., cradle-to-gate, cradle-to-grave, cradle-to-cradle)?

(A) Functional Unit and Product System

A functional unit is a quantified reference that describes the function of the product or system and serves as the basis for calculating impact assessments. For example, *100 MWh of electricity* can be considered the functional unit for evaluating a power plant.

A product system refers to the complete set of processes required to fulfill a specific function. In the above example, the *power plant* is the product system, and its function is the generation of electricity.

(B) System Boundary

The system boundary defines which parts of the product life cycle are included in the study — such as mining, transportation, or recycling — and what flows of energy and materials are considered. This is depicted in Figure 10.3. Setting the system boundary helps determine the overall scope of the project and enables a consistent evaluation of the flows, processes, and product systems involved.

10.3.2 *Inventory Analysis*

Once the goal and scope of the study are defined, the next step involves collecting data on all inputs and outputs associated with each process at every stage of the product's life cycle. For example, the inventory includes:

- Natural resource usage, including water, coal, limestone, and other materials used throughout the life cycle of the product or service.
- Emissions and pollutants released into the environment, such as CO_2, SO_2, and chlorofluorocarbons.
- Environmental impacts such as resource depletion and eutrophication.

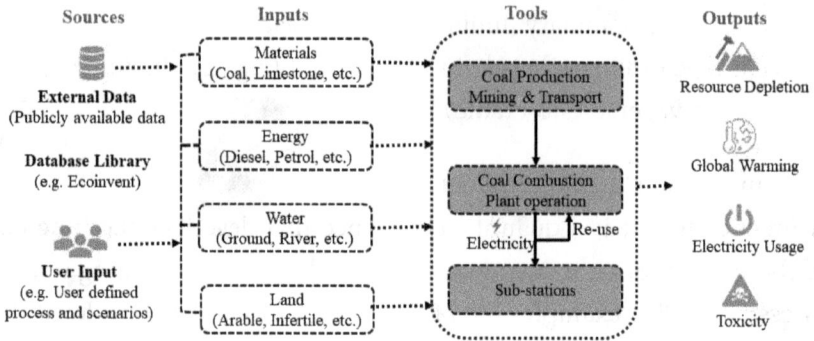

Figure 10.5   Conceptual life cycle inventory of coal-driven power station. Reproduced with permission.[1]

As illustrated in Figure 10.5, an elementary flow diagram is developed in the inventory stage. Inputs for the inventory are typically sourced from secondary data, i.e., public information, software database, occasionally user-specific data, and government reports. Once the input data is collected, output flows are generated. These represent key environmental impacts such as resource depletion, toxicity level, eutrophication, etc. This comprehensive dataset forms the foundation for the next phase of evaluation.

10.3.3 *Impact Assessment*

The third stage of LCA is known as life cycle impact assessment (LCIA). This phase is crucial for understanding how the inputs and outputs from the inventory stage affect the environment, resources, and society. In this step, the outputs from the inventory step are translated into environmental impacts, which are assessed at two different levels:

- Midpoint: Represents specific environmental problems (e.g., climate change, ozone depletion).

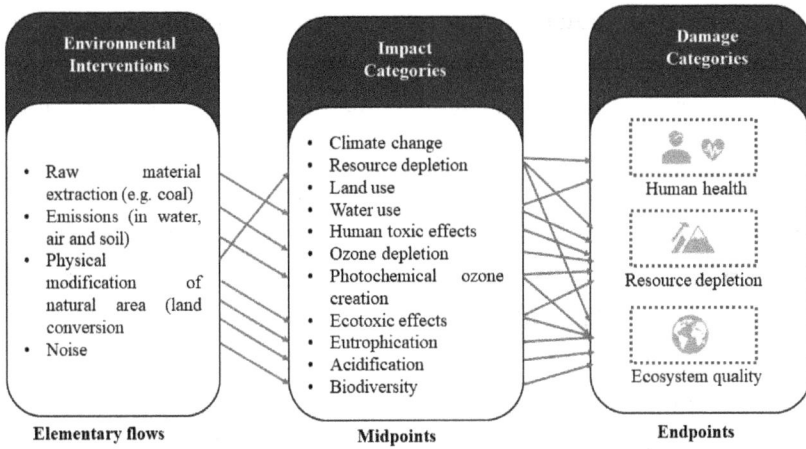

Figure 10.6   Environmental LCIA framework. Reproduced with permission.[1]

- Endpoint: Reflects broader damage to areas of protection (e.g., human health, ecosystem quality, resource availability).

As illustrated in Figure 10.6, the midpoint method assigns inventory flows (inputs and outputs) to impact categories such as such as ozone depletion, eutrophication, smog, etc. The endpoint method, on the other hand, builds upon midpoint indicators to assess long-term and often irreversible impacts, providing insight into the ultimate damage caused to human health, ecosystems, and resources. Some LCIA tools are CML, TRACI, ReCiPe, IMPACT which follow the midpoint method, while ReCiPe and IMPACT also follow the endpoint method. Further, LCIA is performed in four steps: (i) characterization, (ii) normalization, (iii) aggregation, and (iv) weighting.

10.3.4 *Interpretation*

This fourth and final step involves analyzing and converting all the results obtained from previous steps into meaningful and relevant information for the intended audience. The interpretation depends on the initial goal and scope of the study. This step is significant for defining, quantifying, reviewing, and assessing information from the outcomes of the inventory output and LCIA. One can then draw final conclusions from identifying environmental hotspots and evaluating the overall impact of a product, process, or system. Based on the outcomes, this information can support product improvement, process optimization, policy formulation, and strategic planning.

Moreover, based on the scope and boundary, LCA studies can follow different system approaches:

- Cradle-to-Gate: Includes only the stages from material extraction to manufacturing. It excludes the product's use phase and end-of-life.
- Cradle-to-Grave: Covers all five life cycle stages — from raw material extraction to the product's disposal or end-of-life.
- Cradle-to-Cradle: Represents a circular economy model where the end-of-life of a product leads to the creation of new raw materials. Waste is recycled or reused, forming a continuous loop from stage 1 back to stage 1 (Figure 10.7).

Each of these approaches provides different insights depending on the sustainability goals and the decision-making needs of stakeholders.

10.4 Environmental Social Governance

ESG stands for environmental, social, and governance. It is a framework used to evaluate companies based on their environmental impact, social responsibility, and corporate governance — alongside their financial performance. The interconnected roles of each ESG component are shown in Figure 10.8.

The environmental component focuses on how a company manages its consumption of energy, materials, and natural resources, as well as its

FEWER RAW MATERIALS ARE USED

Improved, cost-efficient collection and treatment systems will lead to fewer and fewer materials ending up in landfill and support the economics of circular design.

Products and packaging are designed to last longer and be more durable, using more sustainable materials that can be easily recycled at end-of-life.

RECYCLE

DESIGN

CIRCULAR ECONOMY

Government leadership, producer responsibility, and consumer education and awareness will enable market mechanisms that drive higher resource productivity, innovation and economic growth

Producers are fully responsible for recovering materials from their products and packaging throughout their lifecycle.

REUSE/ REPAIR

PRODUCE

Businesses collaborate and coordinate across sectors to reduce greenhouse gas production and fossil fuel use.

CONSUMER USE

DISTRIBUTE

There are many ways consumers can contribute to a circular economy, like making greener buying choices, sharing assets (e.g. cars, tools) and repairing them or offering them to others for reuse and refurbishing.

Retailers offer products that can be easily reused and refurbished, offer end-of-life take back or maintenance and repair services, and support producers in providing education and awareness to consumers.

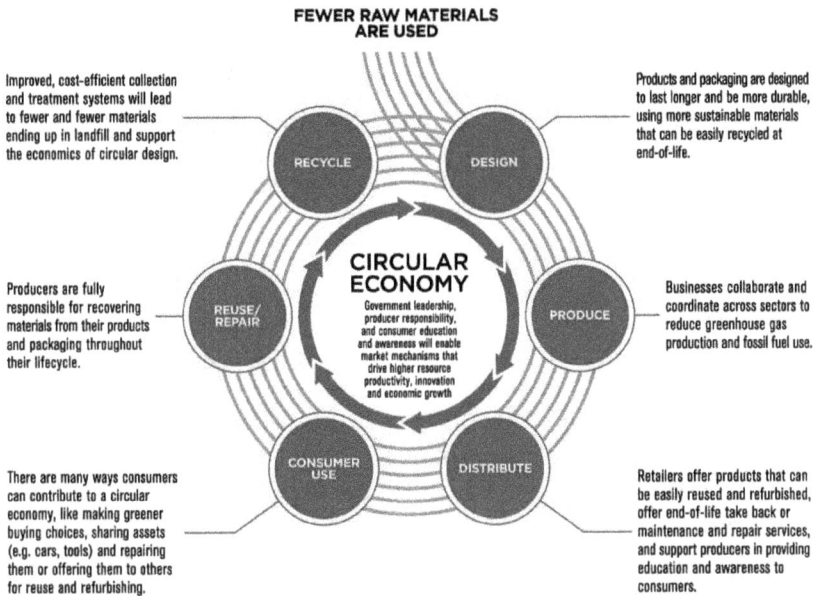

Figure 10.7   Cradle-to-cradle circular economy system. Reproduced with permission.[2]

waste generation and overall environmental footprint. It includes metrics such as carbon emissions and the company's vulnerability or response to climate change. The social dimension reflects the company's values, ethics, and relationships with employees, customers, and the broader community. It covers aspects such as labor rights, diversity and inclusion, and contributions to societal well-being. Governance in ESG refers to the internal systems, policies, and decision-making processes of the company. It ensures compliance with laws and regulations, ethical leadership, transparency, and accountability to stakeholders.

The concept of ESG originated during the post-industrial era. Between the 1950s and 1970s, the global economy was recovering from the damage caused by World War II. A significant labor force was needed to rebuild economies, which led to the formation of mutual agreements between employers and employees to reduce labor exploitation and improve working conditions. As a result, employee welfare and labor

Social

- Employee welfare
- Animal welfare
- Good community relations
- Protecting the interests of shareholders

Environmental

ESG Nexus

Governance

- Carbon emissions
- Climate change
- Ecosystem pollution
- Resource depletion

- Board structure/functions
- Shareowner rights
- Compensation policies
- Bribery and corruption

Figure 10.8 Illustration of the ESG nexus from a sustainable business perspective. Reproduced with permission.[1]

rights were gradually integrated into business policies and corporate governance structures. The formalization of ESG principles began in 2004, as depicted in Figure 10.9.

While many companies in developed nations have since adopted social and ethical standards in their operations, the rapid industrial expansion has also led to severe environmental and social disasters. Tragic events such as the Chernobyl nuclear disaster in Ukraine, the Bhopal gas tragedy in India, and the Exxon Valdez oil spill in Alaska, USA underscore the importance of integrating ESG practices to prevent negligence and ensure accountability. The ESG framework promotes transparency in corporate sectors and plays a critical role in building trustworthy

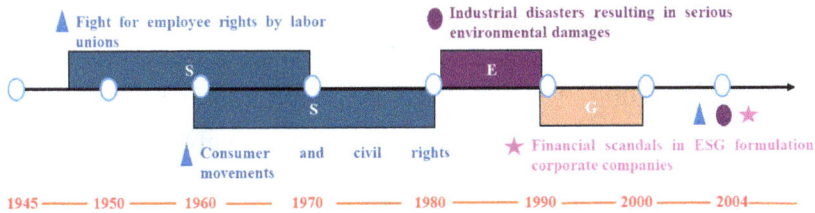

Figure 10.9   Timeline and evolution of ESG. Reproduced with permission.[1]

employer-employee relationships, as well as maintaining sustainable partnerships across industries.

10.5 Industrial Symbiosis

Industrial symbiosis refers to a collaborative approach where industries work together to create synergy by utilizing the waste or by-products of one industry as raw materials for another. According to the Waste and Resources Action Programme (WRAP) in the UK, industrial symbiosis involves the efficient flow and exchange of waste materials within a network of industrial businesses. In broader terms, industrial symbiosis is defined as "*a community of businesses that cooperate with the local community to efficiently share resources (information, materials, water, energy, infrastructure, and natural habitat), leading to economic gains, improvements in environmental quality, and equitable enhancement of human resources for the business and local community*".

There are five opportunities for secondary resource exchange within an industrial symbiotic system:

1. Reduced costs of raw materials and waste disposal
2. By-product exchange and revenue generation
3. Minimization of landfill wasteand carbon emissions
4. Opening for new businesses or startups
5. Enhanced environmental profiles and sustainability credentials

Industrial symbiosis often takes shape within industrial parks, where the co-location of businesses enables efficient material and resource sharing. This model supports job creation and promotes sustainable economic development. Businesses that engage in industrial symbiosis benefit from lower landfill and disposal costs, additional income through by-product utilization, reduced operational expenses, and a lower environmental footprint. In essence, industrial symbiosis turns waste into value while contributing to both economic and environmental sustainability.

10.6 An Example of Life Cycle Thinking and Circularity: The Textile Industry

Globally, the textile industry is one of the most important productive sectors due to its immense contribution to the fashion sector and the economies of developed and developing countries. Like many other industries, the textile sector largely follows a linear economy model and struggles to address the challenges of overconsumption and excessive waste generation. These issues lead to serious environmental and economic impacts, threatening the sustainability of the industry.[3] In this context, implementing LCT in conjunction with a circular economy approach can offer viable solutions to the problems caused by fast-changing fashion trends. The integration of LCT and circularity across different stages of textile production is illustrated in Figure 10.10, promoting sustainable manufacturing and responsible resource consumption.

The goal of LCT analysis is to evaluate and ensure green certification at each stage of the textile product's life cycle, along with appropriate environmental labeling that reflects its environmental footprint. This involves incorporating principles such as rethinking, eco-design, eco-innovation, reuse, recycling, and resource recovery into processes that typically generate waste. Furthermore, circularity analysis enables the transformation of waste materials into valuable resources, supporting

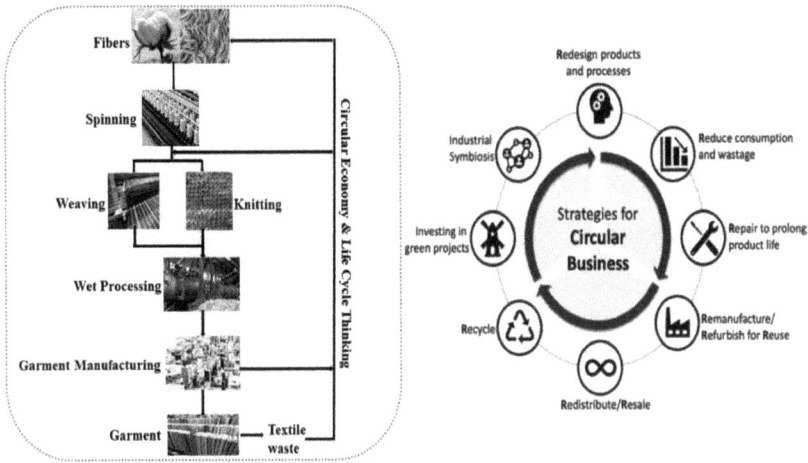

Figure 10.10 A schematic view of textile manufacturing using life cycle thinking and circular economy. Reproduced with permission.[4]

the industry's shift towards a regenerative and closed-loop system. Thus, the combined application of LCT and the circular economy model can pave the way for enhanced sustainability across all industrial sectors, particularly within the textile and fashion industries.

References

1. Selvan, T. and Ramakrishna, S. (2022) *Sustainability for Beginners: Introduction and Business Prospects*, World Scientific.

2. https://www.ontario.ca/page/strategy-waste-free-ontario-building-circular-economy

3. Teixeira, T. G. B., de Medeiros, J. F., Kolling, C., Ribeiro, J. L. D. and Morea, D. (2023) Redesign in the textile industry: Proposal of a methodology for the insertion of circular thinking in product development processes, Journal of Cleaner Production, 397, 136588.

4. Patil, R. A. and Ramakrishna, S. (2023) *Circularity Assessment: Macro to Nano*, Springer.

INDEX